Swarming Landscapes

ADVANCES IN GLOBAL CHANGE RESEARCH

VOLUME 48

Editor-in-Chief

Martin Beniston, *University of Geneva, Switzerland*

Editorial Advisory Board

B. Allen-Diaz, *Department ESPM-Ecosystem Sciences, University of California, Berkeley, CA, USA.*
R.S. Bradley, *Department of Geosciences, University of Massachusetts, Amherst, MA, USA.*
W. Cramer, *Earth System Analysis, Potsdam Institute for Climate Impact Research, Potsdam, Germany.*
H.F. Diaz, *Climate Diagnostics Center, Oceanic and Atmospheric Research, NOAA, Boulder, CO, USA.*
S. Erkman, *Institute for Communication and Analysis of Science and Technology – ICAST, Geneva, Switzerland.*
R. García Herrera, *Facultad de Físicas, Universidad Complutense, Madrid, Spain*
M. Lal, *Centre for Atmospheric Sciences, Indian Institute of Technology, New Delhi, India.*
U. Luterbacher, *The Graduate Institute of International Studies, University of Geneva, Geneva, Switzerland.*
I. Noble, *CRC for Greenhouse Accounting and Research School of Biological Sciences, Australian National University, Canberra, Australia.*
L. Tessier, *Institut Mediterranéen d'Ecologie et Paléoécologie, Marseille, France.*
F. Toth, *International Institute for Applied Systems Analysis, Laxenburg, Austria.*
M.M. Verstraete, *Institute for Environment and Sustainability, EC Joint Research Centre, Ispra (VA), Italy.*

For further volumes:
http://www.springer.com/series/5588

Rob Roggema
Editor

Swarming Landscapes

The Art of Designing For Climate Adaptation

Editor
Rob Roggema
Canterbury, VIC, Australia

ISSN 1574-0919
ISBN 978-94-007-4377-9   ISBN 978-94-007-4378-6 (eBook)
DOI 10.1007/978-94-007-4378-6
Springer Dordrecht Heidelberg New York London

Library of Congress Control Number: 2012953927

© Springer Science+Business Media Dordrecht 2012
This work is subject to copyright. All rights are reserved by the Publisher, whether the whole or part of the material is concerned, specifically the rights of translation, reprinting, reuse of illustrations, recitation, broadcasting, reproduction on microfilms or in any other physical way, and transmission or information storage and retrieval, electronic adaptation, computer software, or by similar or dissimilar methodology now known or hereafter developed. Exempted from this legal reservation are brief excerpts in connection with reviews or scholarly analysis or material supplied specifically for the purpose of being entered and executed on a computer system, for exclusive use by the purchaser of the work. Duplication of this publication or parts thereof is permitted only under the provisions of the Copyright Law of the Publisher's location, in its current version, and permission for use must always be obtained from Springer. Permissions for use may be obtained through RightsLink at the Copyright Clearance Center. Violations are liable to prosecution under the respective Copyright Law.
The use of general descriptive names, registered names, trademarks, service marks, etc. in this publication does not imply, even in the absence of a specific statement, that such names are exempt from the relevant protective laws and regulations and therefore free for general use.
While the advice and information in this book are believed to be true and accurate at the date of publication, neither the authors nor the editors nor the publisher can accept any legal responsibility for any errors or omissions that may be made. The publisher makes no warranty, express or implied, with respect to the material contained herein.

Cover image title: Te Papa, Wellington, New Zealand © Rob Roggema

Printed on acid-free paper

Springer is part of Springer Science+Business Media (www.springer.com)

Foreword

In times of crisis, whether it is economical, climatic or something else, the predominant response is to stop change, kill innovation and stick to well-known procedures and habits. Take the example of consumer trust. When this is going down, people start to postpone investments, which leads to reduced sales, which slows down the economy, which decreases consumer trust, which …, etcetera. The only solution to escape this downturn is, against all odds, to start investing when consumer trust is decreasing. In the words of Einstein[1]: "Any intelligent fool can make things bigger, more complex, and more violent. It takes a touch of genius – and a lot of courage – to move in the opposite direction." When things become complicated or problematic, we should turn against our intuition, which tends to deal with uncertainty by avoiding risk. The same is true in dealing with the impacts of climate change. When uncertainty about the impacts is increasing, at least in the media, the response must not be to build stronger defences 'to keep the danger out'. Instead, the fact that the future is not predictable could offer the freedom to create the future we desire. It should appeal to our fantasy to invent a new kind of thinking, which is capable of solving the uncertainty problem. Or as, again, Einstein formulates it: "We can't solve problems by using the same kind of thinking we used when we created them."

With this in mind, the book *Swarming Landscapes* must be seen as a quest to find the way of thinking that can support spatial designers and planners to formulate spatial responses to the current question how to prepare our society, our cities and landscapes, for the impacts of climate change, given the fact that we cannot predict those.

The way many spatial planning practices currently approach this problem is, paraphrasing Einstein, by using the same kind of thinking to solving the problem, which was used while creating it. Our cities and landscapes were built to serve consuming man, using resources, such as water and energy to the limit whilst returning the garbage to atmosphere, water system and soil. This system (or, the way we live), consisting of sprawling cities, endless car networks and a food chain that ends in ever growing supermarkets, is one of the causes of climate change. We cannot

---

[1] http://rescomp.stanford.edu/~cheshire/EinsteinQuotes.html

expect this system to produce the solutions if the thinking that lies behind its origins stays the same. In Chap. 1 the difficulties of designing for climate adaptation are explored and set the scene for the rest of the book. The core problem of spatial planning practice in dealing with climate adaptation is its tame, quantifiable and fixed character, while climate adaptation is qualified as a wicked, complex system problem. This problematic relationship between adaptation and planning gives reason to explore the quest for a new kind of thinking. In Chap. 2 this exploration starts with positioning the current timeframe in historical perspective. The understanding of uncertainty and turbulence of the current timeframe is the subject of this chapter. The quest continues in Chap. 3 with the way complexity theory can be of use in current planning and design practice, dealing with climate adaptation. The fundamentals of another way of thinking are explored in Chap. 4, where transition and transformation concepts are used to support the change from the current to a future spatial system. In Chap. 5 the role networks can play in intervening and starting up processes of change in the landscape is further elaborated. Networks are seen as better capable of enhancing change in comparison with functions occupying areas. These four chapters form the basis for developing a new planning theory, which is described in detail in Chap. 6: Swarm Planning. This theory based on complexity science and dealing with the spatial system as a complex adaptive system offers better chances to deal with an uncertain problem and a future that cannot be predicted. In Chap. 7 the theory is made suitable to use in planning practice with a description of the Swarm Planning methodology, both in the way the content can be derived as well as how a planning process can be organised. Chapter 8 then follows up with many examples of Swarm Planning, originating from the Netherlands and Australia and dealing with floods, sea level rise, droughts and bushfires. These examples illustrate that the Swarm Planning theory and methodology can be used at different abstraction levels: strategically, identifying single interventions, designing climate landscapes and the design of swarming landscapes. The penultimate chapter, in which the city is framed as an organism and follows the rules of a dynamic system, brings several aspects of former chapters together. In Chap. 10, a future perspective is sketched, in which climate adaptation is the driver for inventing new landscapes and uncertainty is bypassed.

The main conclusion that can be drawn is that Swarming Landscapes are the possible answer when a new kind of thinking is required. It offers a new perspective on designing for climate adaptation and finds solutions to overcome the lethargy of thinking in fear and risk. However, the design of Swarming Landscapes is not easy and is for certain not a trick, which can be learned, copied and used to solve all our imaginable problems. It requires understanding of the specific complexity of the site, its typical problems and the available potentials for innovative thinking.

Melbourne   Rob Roggema

Preface

As Chair of the International Scientific Committee of the World Sustainable Building Conference in 2008 (SB08) I had the pleasure of presenting Rob Roggema and Andy van den Dobblesteen with their award of Best Conference Paper for *Swarm planning: Development of a new planning paradigm which improves the capacity of regional spatial systems to adapt to climate change*.

Four years later it is a bonus to be able to write this preface to the book Rob Roggema has assembled to showcase an expansion of thinking and practice in this new approach to planning. Many of the leading exponents of this planning mode are Dutch. They include those in this volume, their colleagues at the Dutch Research Institute for Transitions, as well as an increasing number now holding academic posts in overseas universities – aiding in the diffusion of this innovation in planning and practice.

In attempting to encapsulate and communicate what I have found to be the essential features of *Swarming Landscapes* I found it useful to deconstruct the book into three elements: the *situation* (problem), the *complication* and the *solution* – the fundamental structure of a research presentation.

The *situation* in question is climate change – classed as a 'wicked' planning problem (Rittell and Webber). It has also been referred to as a 'diabolical' problem by Garnaut in his Australian Climate Change Reports. This is due to the fact that it refers to a *global* process that all nations are fuelling with their greenhouse gas emissions, with no international agreement having been reached to date on mitigation solutions for this slow burn issue (see the boiling frog at http://www.youtube.com/watch?v=TyBKz1wdK0M). Yet the *local* manifestations of climate change are beginning to emerge and will intensify in terms of impact. The urban planning solutions for both mitigation and adaptation challenges are largely *terra incognita* – unknown territory.

Therein lies the *complication*: future≠past. Contemporary planning theory and practice is still rooted in modes of thinking and action dominated by those regimes (in government and industry) that have actually produced the current situation. These regimes tend to be risk-averse rather than innovation-oriented, with the necessary longer-term urban planning horizons hamstrung by short-term (3–4 years) political cycles and an absence of bipartisanship on the issue of city planning. The result is incremental change in a period of urban history when change needs to be transformational.

Which leads us to the search for a *solution*: how to achieve transformation to a new form of sustainable urban development or redevelopment that is resilient in the face of climate change, as well as a raft of other pressures. Fundamental structural changes are required – to the economy, to systems of governance, to societal values and to human settlements.

Roggema advances *swarm planning* as a new planning theory with associated methods capable of enabling radical change to occur. *Swarm planning theory* is linked with the principles of *biomimicry*, complex systems and socio-technical systems, among others, with a view to application in the realm of spatial planning. Of particular interest is the creation of a new process capable of envisioning and implementing planning interventions that can 'flip' or transform significant parts of an urban system, with the prospect of broader-based changes ensuing in a self-organising manner – analogous to a swarm of birds or insects that change their shape suddenly as a result of some particular impulse or signal. *Swarm planning methodology* is concerned with how to develop spatial plans that are able to deal with the somewhat unpredictable local 'landscape' impacts associated with climate change – and are capable of being implemented within a relatively short time frame. A five-layer spatial information model is proposed for the *context* (information) that is required for swarm planning. A charette-style engagement method is outlined as the *process* component.

In summary, I would contrast traditional versus swarm planning as follows:

Traditional planning:

top-down (elites) → impose plan → community resistance (slow or no progress)

Swarm planning:

multi-level (multi-actor) → engagement → consensus plan (implementable)

This book represents an important attempt to chart an alternative pathway for twenty-first century urban spatial planning.

Dr. Peter W. Newton
Research Professor in
Sustainable Urbanism
The Swinburne Institute for
Social Research Swinburne
University of Technology,
Melbourne

Contents

1 **The Difficulties to Design for Climate Adaptation** 1
   Rob Roggema

2 **Turbulence and Uncertainty** 25
   Rob Roggema

3 **Complexity Theory, Spatial Planning and Adaptation to Climate Change** 43
   Wim Timmermans, Francisco Ónega López, and Rob Roggema

4 **Transition and Transformation** 67
   Rob Roggema, Tim Vermeend, and Wim Timmermans

5 **Networks as the Driving Force for Climate Design** 91
   Rob Roggema and Sven Stremke

6 **Swarm Planning Theory** 117
   Rob Roggema

7 **Swarm Planning Methodology** 141
   Rob Roggema

8 **Swarming Landscapes** 167
   Rob Roggema

9 **Cities as Organisms** 195
   Andy van den Dobbelsteen, Greg Keeffe, Nico Tillie, and Rob Roggema

10 **The Best City?** 207
   Rob Roggema

Subject Index 259

About the Author

Dr. ir. R.E. (Rob) Roggema (1964) is Landscape Architect (Wageningen, 1990; PhD: Delft/Wageningen, 2012) and is a design-expert on the issues of adaptation to climate change, sustainable energy supply, climate adaptive ecology and sustainable development. He developed and used this expertise while holding positions at several universities (Delft University of Technology, Faculty of Architecture; Wageningen University and Research Centre; RMIT University, Swinburne University of Technology, The Swinburne Institute for Social Research), governmental organisations (State and Municipal) and Planning and Design consultancies.

From 2003 to 2010 he worked for the province of Groningen on strategic questions and complex projects in the field of sustainability and spatial planning and design. Between 2006 and 2009 he was member of the Sino-Dutch expert team advising a series of Chinese urban development sites on Ecological Building. Between September 2010 and April 2011 he has been appointed as honorary visiting fellow at the Victorian Centre for Climate Change Adaptation Research (VCCCAR) in Melbourne.

Among his special interests are spatial designs of a sustainable energy supply and climate adaptation.

He has written two books on adaptation to climate change and spatial planning and design: 'Tegenhouden of Meebewegen' (WEKA, 2008) and 'Adaptation to Climate Change: A Spatial Challenge' (Springer, 2009).

He developed the Swarm Planning design-concept, which emphasises future uncertainties in current planning strategies, for which he received the scientific award for the best theoretical paper at the Sustainable Building conference (SB08) in Melbourne. In 2012 he received his PhD for his research of: "Swarm Planning: The Development of a Planning Methodology to Deal with Climate Adaptation". In 2009, he received an honorary mention for his entry 'SFBDNA' in the Rising Tides Competition, San Francisco Bay Area.

He currently holds his own consultancy, which advises on sustainable and climate proof planning and design (Citta Ideale; www.cittaideale.eu, rob@cittaideale.eu) and is appointed as Senior Research Fellow at the Swinburne Institute for Social Research, Swinburne University of Technology, where his research focuses on the development of climate adaptive and low carbon urban, metropolitan and regional landscapes.

Contributors

Greg Keeffe Sustainable Architecture, School of Planning Architecture and Civil Engineering, Queens University Belfast, Belfast, UK

Leeds Metropolitan University, Leeds, UK

Francisco Ónega López LaboraTe – Universidade de Santiago de Compostela, Lugo, Spain

Rob Roggema The Swinburne Institute for Social Research, Swinburne University of Technology, Hawthorn, VIC, Australia

Delft University of Technology, Delft, The Netherlands

Wageningen University, Wageningen, The Netherlands

Sven Stremke Landscape Architecture, Wageningen University and Research Centre, Wageningen, The Netherlands

Nico Tillie Faculty of Architecture, Delft University of Technology, Delft, The Netherlands

Wim Timmermans Groene Leefomgeving van Steden, Wageningen UR – Van Hall Larenstein – Tuin en Landschapsinrichting, Velp, The Netherlands

Andy van den Dobbelsteen Climate Design & Sustainability, Faculty of Architecture, Delft University of Technology, Delft, The Netherlands

Tim Vermeend UC Architects, Groningen, The Netherlands

Introduction

In many parts of the World the topic of Sustainable Energy Policy is seen as part of the Economic agenda. On the one hand side this might be beneficial, because once energy measures become part of an economic program they will get executed. The downside to this is that once the economic growth decreases, energy is the first part that drops of the agenda. The danger of linking sustainability issues to the economic policy field can be seen in the debate around the introduction of a carbon price in Australia. The discussion is no longer about the, in itself valuable, goal to minimise carbon emissions, but about the economic (read: money) pros and cons. What sustainable energy policy is suffering from is, in a slightly different way, also true for climate adaptation. In this case it is not a purely economical background that hinders realisation of important goals, but the 'calculator illness'. Climate change impacts are only valid if they can be calculated in terms of how many centimetres sea level rise, the exact chance a flood or bushfire will happen and the amount of economic assets or peoples lives at risk. Climate change is seen as a threat and the risks, vulnerabilities and disasters must be assessed. However, no assessment ever stopped a disaster from happening. Even worse, most of the assessments do not include the magnitude of a real disaster and therefore calculate a phantom safety.

In current practice the valve of climate change is often positioned towards the '*mathematisation*' half of the problem. The other half, where climate change is seen as a chance, is often neglected or undervalued. Here the creativity of finding new solutions and new ways of thinking can be explored in order to find solutions for problems that the old way of thinking will never be capable of. Climate adaptation is positioned as a major factor in determining the future way of living and not only as a singular environmental, 'green' issue. In order for us to open the valve towards the creative half and for creativity to get challenged, the complexity of the problem needs to be investigated. This book opens the valve and explores uncertainty, turbulence, complexity, transformation and networks a s elements to create design with, allowing the landscape to behave as a swarm: Swarming Landscapes emerge.

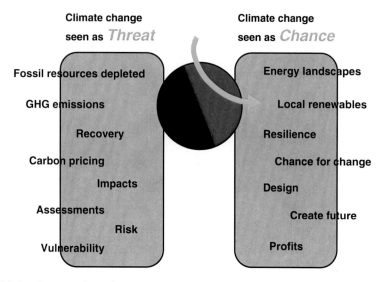

This book opens the valve.

Chapter 1
The Difficulties to Design for Climate Adaptation

Rob Roggema

Contents

1.1	Introduction	2
1.2	Design and Policy	2
1.3	Current Planning Practice	5
	1.3.1 Melbourne Metropolitan Area	6
	1.3.2 Greater Groningen Area	8
	1.3.3 Prevailing Plan Versus Climate Adaptation Requirements	12
1.4	The Spatial System as a Complex Adaptive System	15
1.5	Wicked Problems	18
1.6	Conclusion	21
References		22

Abstract In this chapter current spatial planning practices are analysed and it is concluded that these practices are not yet suitable to include wicked problems. The topic of climate change adaptation is seen as a wicked problem, meaning that the problem has no definite solution, the problem is essentially unique, while 'solving' the problem new questions can be asked and new information can be added and the solution on one level of the system may lead to problems at the next. Therefore, a fixed planning solution for an urban region or landscape is the ultimate failure in dealing with such problems. Current planning needs to be adjusted in order to be better capable in dealing with climate change adaptation problems. In the planning process and in the results of planning more room for unexpected, unprecedented impacts and new knowledge needs to be created.

Keywords Spatial planning • Wicked problem • Climate change • Design

---

R. Roggema (✉)
The Swinburne Institute for Social Research, Swinburne University of Technology,
PO Box 218, Hawthorn, VIC 3122, Australia
e-mail: rob@cittaideale.eu

1.1 Introduction

Climate change may rejoice large attention over the last decade. The work committed by the Intergovernmental Panel on Climate Change, the United Nations Climate Change Conference in Copenhagen (COP 15) and several striking disasters, which were related to climate change, all lead to increased media coverage of the subject. Almost equal attention has been given to the debate whether climate change is real, climate-sceptics and general disbelief. Climate change is facing difficulties to getting the message across. Still, subjects, such as the economy, health, education, taxes or income policy are found more important.

Spatial planning and design potentially forms a very suitable platform for climate change. It aims to prepare society for the longer term and to provide a good and safe living environment. These properties that are linked with spatial planning seem to fit perfectly with a topic such as climate change. The aim to think on the longer term and to protect society against and prepare it for the impacts of climate change is in line with the character and the impacts of climate change on society.

However, *malintegration* of climate change in spatial planning is often the case. Policies regarding climate change are often limited to quantitative ambitions about to what level emissions must be reduced, how much energy must be saved or how high a price on carbon is reasonable. The relation with the making of spatial plans is often reduced to the phase of the analysis, in which risk or vulnerability assessments are carried out and the bushfire and flood prone areas are defined.

In this chapter we are looking into elements that cause this *malintegration*. We will start with framing design in relation to policymaking, followed by an analysis of current planning and to what results this leads, taking the Melbourne Metropolitan Area and Greater Groningen Area as examples. The second part of this chapter looks into spatial systems as complex adaptive systems, climate change as a wicked problem and the misfit this causes with current spatial planning.

1.2 Design and Policy

The role of planning and design in relation to policymaking can be framed in different ways. Massoud Amin and Horowitz (2007) distinguish three fundamentally different types (Fig. 1.1):

(A) Policy determines design (top-down). In this type of approach the policy 'decision' to develop a certain system is most important. Once this policy-decision has been made the system can be designed and realised. In general these *large-scale systems* take a long time to realise, are expensive – and must therefore last a long time, change at a slow pace and are difficult to modify once they are realised. Despite these systems are highly reliable and assure, these benefits also imply higher risks. The suggested assurances are not that assured. If one component in these large and complex systems fails the entire system breaks

1 The Difficulties to Design for Climate Adaptation

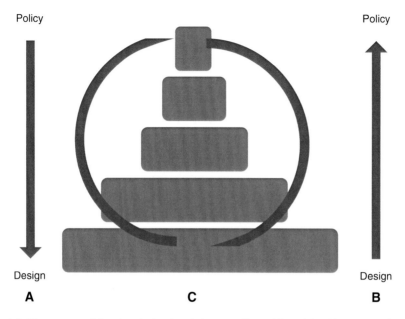

Fig. 1.1 Three ways of framing design in relation to policymaking (After Massoud Amin and Horowitz 2007)

down. These systems do not learn from earlier use or failure and the system objectives are not refined on the basis of actual experiences. A good example of these large-scale systems is a power plant. The framing of design as the engineer emphasises fixed and unchangeable solutions. The designers' role is to make the desired system function or a reality. It supposes that the World is 'makeable', the only challenge is to solve the technological problem. The top-down approach strives to create order and abandon chaos and unpredictability;

(B) Design determines policy (bottom-up). This type sees the design as the driving factor for development of systems. Evolution of the system is based on the survival needs of the elements (or components). The resulting system is created through the interdependencies between these survival objectives and the assurance is derived from adaptation resulting from the interaction of elements. No controlling factor, interactions determine performance of the system, including the emergent properties. These *complex large-scale systems* are developed through a distributed process, in which separate entities, each with their own objectives, risks and individual motivation without an overall integrated schedule or plan with fixed start or end, contribute parallel to a shared system concept and interacting at a fast pace. These interactions in themselves do not provide an assured system (e.g. reliable, short response times, secure), but the agility is very high and newly created technology can be rapid inserted into use. New functions and system capabilities emerge out of integrated contributions of the entities (emergent properties). Good examples of these systems are the Internet and e-business. In these randomly unplanned systems order emerges out of chaos. There is no 'hidden-hand' directing the systems direction, but the system behaves completely flexible under perpetuating change. The 'design' of the system emerges out of individual entities, which, while randomly interacting, determine the properties of the system;

(C) Design and policy mutually influence each other (interconnectedness). The third type of framing design in relation to policymaking combines the directive with the emergent approach. There is a central directive mechanism, but there is no entity with complete control over multi-scale distributed, highly interactive, networks or to manage them in real time. In these *complex large-scale systems with interventions* innovative changes in the systems' conditions and parameters, cause its components to evolve, leading to changes in components' (re-)actions and decisions to 'tinker' with systems rules and structure. Some components increase the level of tinkering and develop strategies to relieve stress, while others fail and shrink or die. There is no conventional mathematical method able to handle the complexity and interconnectedness of these systems. However, through the combination of clear interventions and the use of emergent properties of the system, assurance levels are low, but the system pertains a long life. The system is highly reliable on the longer term and has the agility to be adjusted to short-term requirements. Examples of these types of systems are the body or the brain and city planning. The design intervenes in the system causing *directed* emergence, e.g. flexible and changeable but with a certain goal. The design defines the intervention that starts the emergence in a certain desired direction. If top-down systems are referred to as order and bottom-up systems as chaos, these directed emergent systems are seen as truly complex and at the edge of order and chaos, shifting between anarchy and stagnation, the place where a complex system can be spontaneous, adaptive and alive (Mitchell Waldrop 1992).

If we link the different ways to frame design to current planning practice, type A in its core is how planning is currently practiced. The societal demands are quantified and subsequently executed. Design is virtually absent and reduced to engineering the spatial layout. Type B reflects a random, chaotic development, without a clear direction or goal. Due to the properties of its components emergent patterns auto-develop, such as we for instance can witness in slums. Type C offers the opportunity to enforce change in a desired direction, through an intervention in combination with emergent property of complex systems. Being in the twilight zone between order and chaos, this approach let the spatial layout swop from anarchy to stagnation and back, letting the system adapt and stay alive. The main direction is clear and design interventions cause developments emerging towards that desired future. Framing design in a way that is develops the right intervention for change in the right direction to occur, makes it also the most complex one. Is type A simple and controlling the system completely, and is type B chaotic with no need for interventions, type C requires understanding of the interconnectedness of its components and the way to improve its agility. It needs to balance straightforward regulatory measures with incentives and low-tech strategies (Russel 2011).

1.3 Current Planning Practice

The way planning and design is currently being executed originates from scientific paradigms of 'to measure is to know' and a strong belief in technological, quantifiable, engineering solutions for problems: approaches that can be framed as type A. Besides this cultural dominated approach of spatial planning and design differences in time-horizons play an important role. Three fundamentally different time-horizons are distinguished (Roggema and Van den Dobbelsteen 2008): long-term developments, such as climate change and energy supply, the planning horizon and the political timeframe (Fig. 1.2).

Climate change and the global energy supply are long-term developments. Starting today – or ongoing for a longer time already, the changes will continue for the next century and beyond. Build houses and generated urban patterns have a similar time span: they also last a century or more. Thus, in theory, it should be easy to combine and integrate these long-term changes in spatial planning practice. However, spatial planning mostly fixes its horizon on a period of maximal 10 years. This relatively short-term focus in a context of predicted long-term changes creates difficulties.

Current spatial planning systems are not very flexible. In the process of making a spatial plan, the requirements every single land-use function has is defined quantitatively in the form of the number of houses, the hectares of land for new industrial areas, the area needed for the ecological system and so on. Subsequently, these programmatic 'volumes' become the components of the spatial plan. The quantitative demands determine the spatial layout in a fairly linear way and once these components are part of the plan the future is fixed. This can be characterised

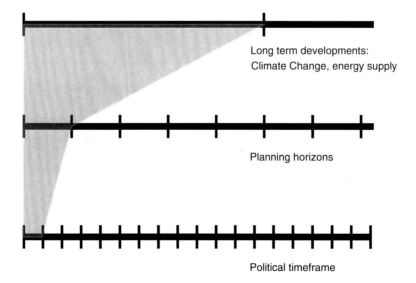

Fig. 1.2 Connection of long- and short-term (Roggema and Van den Dobbelsteen 2008)

as a tame planning approach, in which rationality and standards are dominant. Conklin (2001) characterises tame problems as follows:

- Relatively well-defined and stable problem statement;
- Definite stopping point, i.e. we know when the solution is reached;
- Solution can be objectively evaluated as being right or wrong;
- A problem belongs to a class of problems which can be solved in a similar way;
- Solutions, which can be tried and abandoned.

This approach is used to imagine a spatial layout in a Business-as-Usual scenario for Melbourne Metropolitan Area and Greater Groningen Area respectively. The Australian example illuminates a development scenario, which is based on the dominant role of landownership focusing on individual rights to build and develop, while the Dutch example exemplifies a more commonly developed and negotiation plan-based approach, in which general interests of community play a larger role.

1.3.1 Melbourne Metropolitan Area

The Victorian context for spatial planning and urban development, as reflected in many reports and articles (Department of Infrastructure 2002; DPCD 2008; Moodie et al. 2008; Gleeson 2010; Wheeler 2010) and is dominated by questions how to deal with the implications of population growth, such as housing demand and affordability and infrastructural requirements. The majority of the population growth, from the current four million to an expected seven million people in 2030 (DPCD 2008) will be accommodated in Melbourne Metropolitan Area.

Major planning decisions are taken at the council level and are strongly influenced by the interests of landowners and developers. The State of Victoria intervenes by Ministerial decisions if a certain proposed development at council level is perceived as undesirable. These interventions are often seen at local level as being unwelcome and sudden. The pace of developments, i.e. rapid population growth, does not leave much time for contemplation and reflection about the desired future on the longer term. The repeated extensions of the Urban Growth Boundary (The Age 2010, 2011) can be seen as decisions being made under pressure of immediate housing demand and do not take the real costs of living, such as energy costs for long-distance transportation and electricity use and heating in (large) houses, at the outer fringes of Melbourne Metropolitan Area into account. Current planning decisions, in combination with a traditional car dependency, results in an ongoing, rapid growth of a car-based society in the form of widespread urban sprawl.

After establishment of relatively compact and coherent neighbourhoods in the surrounds of what we now know as the CBD, recent urban developments in the Melbourne Metropolitan Area (Fig. 1.3) have taken place in peri-urban areas. These so-called Greenfield developments start from scratch, without taking into consideration natural landscape-forming elements, such as rivers, topography and

1 The Difficulties to Design for Climate Adaptation 7

Fig. 1.3 Melbourne Metropolitan Area (Source: Google Earth)

Fig. 1.4 Results of Australian urban design (Source: Google Earth and collected from internet, Roggema 2010)

nature in their spatial layout (Presland 2009). These urban developments 'roll over' the underlying landscape, whilst forgetting of the natural drivers.

The key spatial characteristics (Fig. 1.4) in any ordinary development are dominated by two elements: the size of the house and the dominance of the car. It leads to wide roads, a prominent location of the garage in recent architecture (Wheeler 2010) and the overshoot of parking space near shopping malls.

However, developing these Greenfield sites at a large distance from the City does have downsides.

Research by Dodson and Sipe (2006, 2008a, b) demonstrates that these developments are the most vulnerable to rising oil prices and value of mortgages. Moreover, it is not difficult to understand the negative effects, as large houses and commuting takes a lot of energy and valuable land is used extensively for the mono-function of living.

Besides social and environmental benefits densification alongside main public transport axes can financial advantages of up to $300 million per 1,000 living units (Adams 2009).

And last but not least, many of these developments are projected and realised much closer to or inside bushfire or flood prone areas, placing inhabitants at higher risks.

The question what would happen if this Business-as-Usual planning practice continues, e.g. if we assume that big houses are plot-by-plot developed and these are located near existing (larger) villages or country towns and major infrastructural networks and depending land ownership and decisions to extend the Urban Growth Boundary, is answered in the speculative sequence of maps (Fig. 1.5). It demonstrates the unstoppable nature of the process as it is directed on individual decisions and not based on a commonly felt and desired future vision. The process results in a future many people say they do not appreciate, but are unable to change pathways. Despite the fact there is no overarching view about the future, the physical result is highly consistent: low-density urban sprawl with a high volume-area ratio houses (the amount of cubic metres realised at every square metre plot).

1.3.2 Greater Groningen Area

In the Greater Groningen Area (Fig. 1.6), a region in the Northern part of the Netherlands, the tension between a still rapidly growing and economically viable core area and the periphery with a shrinking population and decrease of jobs and amenities, determines the planning task. As restated in the prevailing regional plan for the province of Groningen, the matching of the right amount and types of housing with demand and the division of industrial areas under rapidly changing demographics and increasing differences between sub-regions is the main issue to be dealt with (Provincie Groningen 2009).

Major planning decisions are taken at the municipal level and the majority of the Masterplanning of new neighbourhoods is carried out through municipal offices. Within these Masterplans developers are given the rights to develop (parts of) the plan. The public space and infrastructure is designed, realised and maintained by the local government. There is a strong belief in the planning power and quality of governmental agencies, functioning as the 'herd' of common goods. These Masterplans are made for a large area at once and provide development/building capacity for a longer period. This is the reason large urban entities are designed, developed and built, before the next entity is planned for. The characteristic urban form of each of the developments represents a certain zeitgeist and the moment of building can be determined on the basis of how the area looks like. The cities are built up in concentric circles of subsequent developments.

The first parts of the historic city of Groningen were built at the ultimate (northern) part of the *Hondsrug* ridge, which was surrounded by lower and wetter areas. First living areas were developed south of the central city in the topographical higher

1 The Difficulties to Design for Climate Adaptation 9

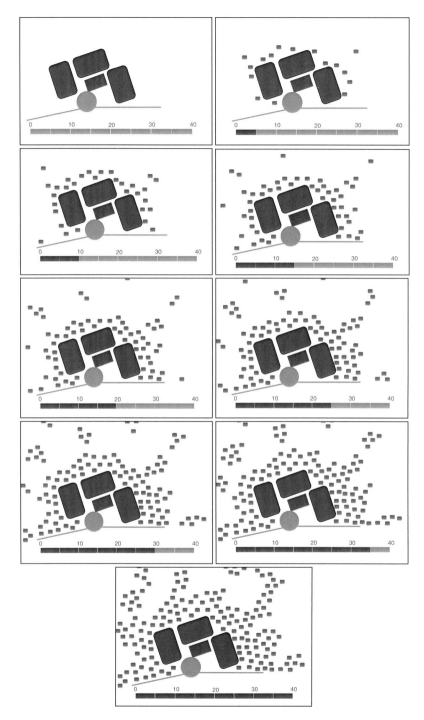

Fig. 1.5 Speculative representation of the development Melbourne Metropolitan Area in 5-year steps (Source: Roggema 2010)

Fig. 1.6 Greater Groningen Area (Source: Google Earth)

Fig. 1.7 Results of Dutch urban design (Source: Google Earth and collected from internet, Roggema 2010)

landscapes. However, since the 1960s new living areas were projected in low-lying landscapes. These neighbourhoods were built through heightening the area with sand with no connection to the underlying water- and natural system.

The key spatial characteristics (Fig. 1.7) of any regular development in the Netherlands are the small and attached houses and relatively compact developments in higher densities of 30 dwellings per hectare or more. Due to the comprehensive development of neighbourhoods, urban patterns reflect a repetitive structure of small streets, small gardens and orderly organised houses, with generally a low volume-size ratio, in a row.

The Dutch planning framework of rezoning every 10 years aims to be able to include or react on changed circumstances. However, because of the municipal Masterplanning developments take place in specific periods of development. In combination with a tradition of tackling problems with solutions of the past, even if

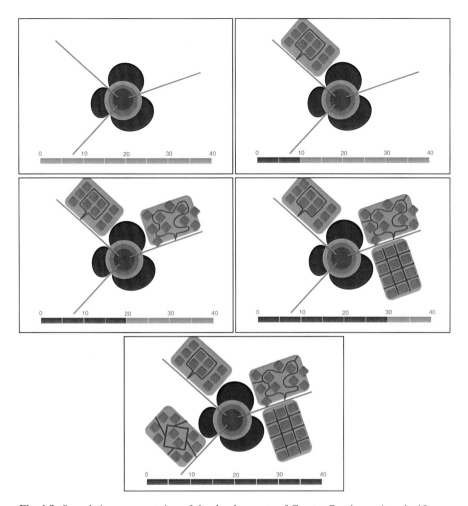

Fig. 1.8 Speculative representation of the developments of Greater Groningen Area in 10-year steps (Source: Roggema 2010)

problems are completely new, leads to a locked-in situation of spatial planning and fixed spatial layouts.

The question what would happen if this directed Business as Usual planning approach continues, e.g. the development of comprehensive designed unchangeable compact urban designs for a larger area, is answered in Fig. 1.8. Historically, Dutch cities have grown slowly and organic, but starting at the beginning of the twentieth century planned developments start to take place on a neighbourhood-by-neighbourhood basis. Each of these plans resembles the zeitgeist of the development. In subsequent jerky phases entire areas are added to the existing urban area.

Both Dutch and Victorian planning frameworks result in spatial models in which unforeseen changes are difficult to implement. In dealing with unprecedented changes, such as (extreme) weather events current planning approaches no longer satisfy.

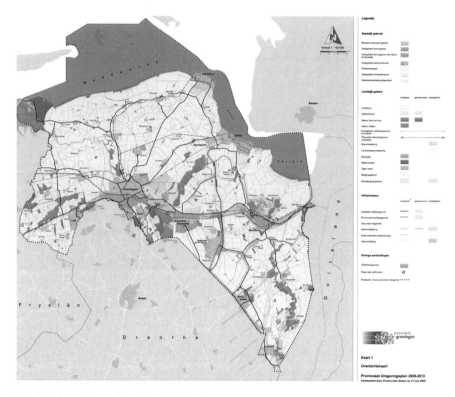

Fig. 1.9 Prevailing regional plan for Groningen Province (Provincie Groningen 2009)

1.3.3 Prevailing Plan Versus Climate Adaptation Requirements

If we take a more in depth look into the possibilities to integrate climate requirements in existing plans, we take the Groningen regional plan as an example. The prevailing plan was set in 2009. In this plan, regional policies for spatial affairs, traffic, water management and the environment are integrated (Provincie Groningen 2009). These policies are brought together on one regional map (Fig. 1.9).

The prevailing regional plan can be characterised as a continuation of an unchanged spatial organisation. The plan does not focus on developing processes neither on the emergence of new structures and patterns. There are only a few adjustments proposed in the plan, such as a corridor alongside the freeway to the east, new living areas to the east and west of the central city and small adjustments to the ecological main structure.

Moreover, if we analyse the three most recent regional plans for the Province of Groningen (2000, 2006, 2009), the changes in the content of aims and policies, the distinguished chapters as well as the accompanying functional maps are marginal. Policies, once defined in the first plan (2000) are to a large extent repeated in the second and third plan. The repetitive character of the three consecutive plans is illustrated through a quantitative analysis of the respective function maps. The size of the area that is allowed to change its function during the period between the first and third plan, has been measured as a percentage of the entire area. It is concluded that a little more than 2% is permitted to change over the effective term of the three plans (e.g. 13 years). Despite the fact that one series of plans do not fully represent all spatial plans that have been made in the Netherlands over the last 15 years, the Groningen example highlights the 'incrementality' of changes. Within the period of 13 years the regional government has the chance to strategically review and change its spatial policy in the light of uncertainty and (un)expected change through increasing the flexibility in the plan to make changes possible (at least), but change is limited to the mentioned 2%. The question is, however, how much change needs to be made possible in the plan in order to deal with (unknown) future climate change. This question has been addressed in one of the first studies on the relation between climate adaptation and spatial planning in the Netherlands (Roggema 2007). In this study climate analyses (KNMI 2006; DHV 2007; Alterra et al. 2008) and the effects on existing functions (MNP 2005) are used to design a spatial layout, which estimates the amount of space that is required to deal with anticipated changes: the 'Idea-map for an adaptive Groningen' (Fig. 1.10). The following elements form the Idea-map:

1. The lowest parts are used to store water. Even if droughts are long lasting, these areas enable to preserve water. It provides a structure to develop a robust ecological connection between the Dollard and Lauwers Lake. Existing brooks discharge their water from the higher Drenthe plateau towards this new storage area. In the ecological zone existing as well as colonising species are able to find suitable habitats;
2. This storage area also functions as the water resource to supply agriculture with enough water of good quality. This is especially relevant in the Peat Colonies, where drought will have largest impacts. The water will be transported making use of the existing canal system in the Colonies. The availability of water is essential for the potato starch production in the area. This supply is necessary from a quantitative point of view, but is also needed for a qualitative reason. In the 'dry' KNMI-scenarios, the groundwater level will drop and might even become saline. Addition of sweet water prevents the available water for agricultural use from becoming useless;
3. Salinity along the northern coastal zone increases, due to sea level rise and increased salty seepage. This makes this area suitable for saline agriculture and aquacultures;
4. Near Lauwers Lake, Dollard and around Delfzijl space is created to inundate water from the sea into so-called 'climate buffers' (Bureau Stroming 2006), allowing a brackish environment to emerge. The combination of salt and fresh water makes it possible to generate energy in an osmosis plant;
5. In front of the Northern Coast new barrier islands are created for protection, to develop nature and to provide development locations for living and recreation;

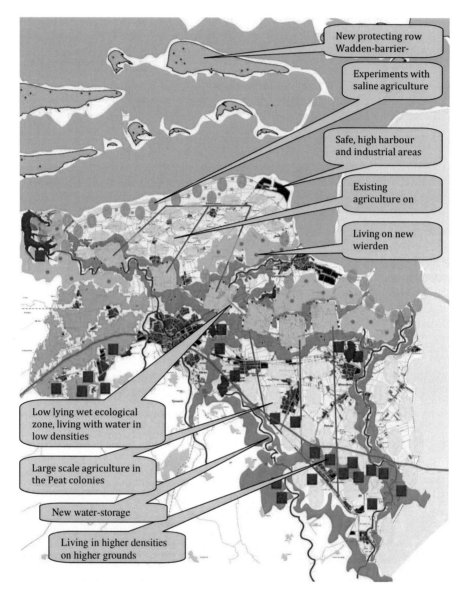

Fig. 1.10 Idea-map climate adapted Groningen (Roggema 2007)

6. The safest parts in the province to create living areas are found where topography is higher: around Leek and the city of Groningen and in the southern part of the Peat Colonies and Westerwolde.

When this design is analysed, approximately 30% of the land area needs to potentially undergo a functional change to eventually adapt to future climate change. The figure of 2% change allowed in the prevailing regional plan is, compared with these results, insufficient.

1.4 The Spatial System as a Complex Adaptive System

The relatively dissatisfaction with current planning methods to include longer term and complex problems, urges for an alternative approach. Complexity theory, on which a more detailed reflection can be found in Chap. 3, may offer a way to deal with complex problems, because these types of problems require continuous evolution, self-organisation and adaptation. But is it plausible to define spatial systems as complex adaptive systems?

A complex system is a system whose properties are not fully explained by an understanding of its complement parts (Gallagher and Appenzeller 1999). These systems are subdivided in Ordered, Critical and Chaotic systems (Lenton and Van Oijen 2002). The Critical ones, dubbed 'the edge of chaos', regard complex adaptive systems (CAS), which consist of three essential elements (Levin 1998): sustained diversity and individuality of components, localised interaction amongst components and autonomous process that selects a subset for replication or enhancement. From these elements the following properties emerge (Holland 1995): continual adaptation, absence of a global controller, hierarchical organisation, generation of perpetual novelty and far from equilibrium dynamics. Complex systems can be defined according their order of dynamics and control (Jørgensen and Straskraba 2000): systems with fixed parameters and structure (class I), with variable parameters and fixed structure (class II), with variable structures and parameters (class III) and systems with a change in goal functions, which determine long-term behaviour of the system (class IV). Lenton and Van Oijen (2002) demonstrate Gaia as a complex adaptive system of class IV, showing properties that can be linked with complex systems, complex adaptive systems, adaptivity and self-organising (Table 1.1).

On the basis of the properties of complex adaptive systems as discussed before, while adding dispersed interaction (Arthur et al. 1997) four general properties of any CAS may help to identify spatial systems as complex adaptive systems (Holland 1995):

1. **Aggregation.** Every system shows non-homogeneities, in which some elements are more similar than the background. Patterns of aggregations and hierarchical organisation are developed as (a) a natural consequence of self-organisation of any complex system (O'Neill et al. 1986; Holling 1992) or (b) an essential element in the later development of the system. The patterns emerge from local interactions through endogenous pattern formation (Levin and Segel 1985; Murray 1989);
2. **Nonlinearity.** Path dependency is a result of nonlinearity. In ecological terms the "early recruitment changes the landscape for future colonists" (Levin 1998). In more general terms an early intervention determines to a large extent the pathway afterwards;
3. **Diversity.** The generation and maintenance of diversity is fundamental to adaptive evolution. Diversity provides resiliency and a hedge against extinction. The key to resilience is the maintenance of heterogeneity, the essential variation that enables adaptation;

Table 1.1 Gaia as complex adaptive system

	Property	Gaia
Complex system	Interwoven parts	
	Properties are not fully explained by understanding the parts	
	Sustained diversity and individuality	Organisms
	Local interactions	
	Autonomous selection process	Natural selection
	Non-local interactions	Global mixing atmosphere and ocean
Complex adaptive system	Far from equilibrium dynamics	Of life, atmosphere and ocean chemistry
	Generation of perpetual novelty	Through evolution
	Hierarchical organisation	Organism, ecosystem, Gaia
	Absence of global controller	
Adaptation	Not evolved by natural selection at whole system level	Various types of selection operate within the system
	Change at system level	Developmental trends, enhancing life
Self-organisation		Evolution, reshaping the system, allowing higher levels of system control
Class III	Changes in parameter and structure of the system	Evolution of responses to prevailing environmental conditions
		Evolution of environment-altering traits
Class IV	Change in goal function	Switch from anaerobic in aerobic conditions, leading to aerobic attractor state, allowing proliferation of new types of life

4. **Flows.** Flows provide interconnections between parts and transform the community from random into an integrated whole. In this transformational process clusters form, flows become modified and the system assumes shape through a process of self-organisation and gives the system its character.

Despite the fact that there is only little evidence found in literature, defining spatial systems as complex adaptive systems, in this chapter it is argued that every spatial system, no matter how its boundaries are defined, can be seen as a complex adaptive system. Following the idea that a smaller part of a certain system contains the same properties as the whole (fractals), every part of Gaia, e.g. every spatial system, spatial system, has identical properties as the whole. Or, the other way around, if we can explain properties of complex adaptive systems in a spatial manner, they can be used to define the spatial system as such. Applying this thought to the four general properties of complex adaptive systems leads to the following proposition:

Aggregation, translated in spatial terms, would mean that hierarchies and differences in a certain area are apparent. When networks of energy, traffic, water consist

1 The Difficulties to Design for Climate Adaptation 17

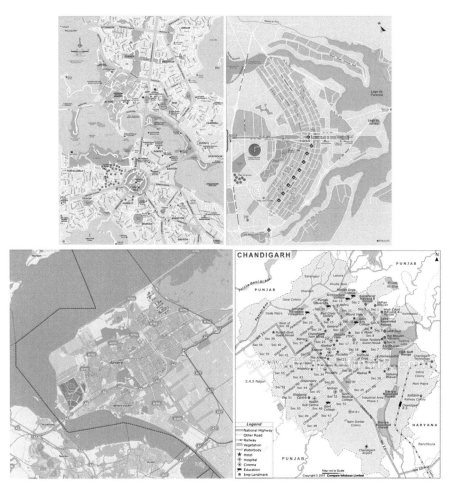

Fig. 1.11 Maps of predetermined, planned, cities: Canberra (AUS), Brasilia (BRA), Almere (NED) and Chandigarh (IND)

of main connections and lesser important ones, when there are core nodes and parts of the networks without significant node the spatial system is non-homogenous. The same property is apparent if there are differences between centres and peripheries, such as between suburban neighbourhoods and the CBD.

Nonlinearity in spatial planning can be seen as an intervention in the landscape, such as to project infrastructure or to develop an activity centre or clustered amenities. The choice to intervene is changing the pathway and determines a path-dependency, which is fundamentally new and cannot be 'undone'. At a higher level, the choice to build New Towns, such as Canberra, Brasilia, Almere or Chandigarh (Fig. 1.11) are similar decisions intervening in developmental pathways of entire countries.

The creation of *diversity* in specific areas aims to increase resilience of these spatial systems. This can be arranged by increasing the spatial and functional heterogeneity. In the city for instance, the diversity of social groups can be enhanced through introduction of a wide typology of housing, a diverse pallet of amenities or different spatial typologies, such as high-density CBD's with large parks or low rise neighbourhoods with a mixture of housing, markets, landmarks and public spaces. The heterogeneity in the landscape can be enhanced in the same way, only the dimensions differ.

In order to let *flows* increase interconnectedness, enhancing self-organisation networks are the key elements in the spatial domain. The intensity, importance and connections of water, energy, transport and communication/social networks determine the capabilities of the system to self-organise and develop its (spatial) character. The role of networks, network nodes and connectedness in spatial design and planning will be further discussed in Chap. 5.

The question remains why defining the properties of complex adaptive systems as spatial elements is important. Apart from the fact that a spatial system as such can be defined as a complex system, focusing on the complex adaptive properties of spatial systems makes it possible to explore spatial development pathways, which are capable to evolve, self-organise and adapt, characteristics that support the system in dealing with wicked problems.

1.5 Wicked Problems

Rittel and Webber coined the term wicked problems in a planning context in their landmark article "Dilemmas in a general theory of planning" (Rittel and Webber 1973). They defined the following ten characteristics:

1. There is no definite formulation of a wicked problem;
2. Wicked problems have no stopping rules;
3. Solutions to wicked problems are not true or false, but better or worse;
4. There is no immediate and ultimate test of a solution to a wicked problem;
5. Every solution to a wicked problem is a "one-shot operation"; because there is no opportunity to learn by trial-and-error, every attempt counts significantly;
6. Wicked problems do not have an enumerable (or an exhaustively describable) set of potential solutions, nor is there a well-described set of permissible operations that may be incorporated in the plan;
7. Every wicked problem is essentially unique;
8. Every wicked problem can be considered to be a symptom of another {wicked} problem;
9. The causes of a wicked problem can be explained in numerous ways. The choice of explanation determines the nature of the problem's resolution;
10. [With wicked problems,] the planner has no right to be wrong.

De Jonge (2009) grouped these ten, into four mutually related themes for the field of Landscape Architecture:

- In design problems the problem definition and solution are inseparable in content and time: once the problem is formulated additional questions can be asked and more information can be asked and provided. The design process is an alternating

sequence of generating variety and reducing variety, searching for possibilities and evaluating.
- Design problems are *social* systems problems: The problem is an interpretation of a problem situation and for every wicked problem there is always more than one explanation, depending on the worldview of those involved. Therefore a solution can never be right or wrong.
- Design problems are systems problems: a solution for a problem in a certain system may be a wrong solution for higher-level systems. A solution can never be definitive tested.
- Every design problem is unique: the uniqueness of every situation requires a new process of argumentation, deliberation and learning of the peculiarities, which makes it impossible to simply copy solutions.

How are these wicked problems related to spatial planning and climate change? As discussed before, spatial systems can be defined as complex adaptive systems. This means that spatial systems are capable of 'producing' wicked problems. Maybe not every problem is a wicked one, but the nature of these systems will cause wicked problems to develop and occur. Moreover, these systems are in essence also suited to deal with them. Secondly, climate change is defined as a wicked problem. *"Climate change* is a pressing and highly complex policy issue involving multiple causal factors and high levels of disagreement about the nature of the problem and the best way to tackle it. The motivation and behaviour of individuals is a key part of the solution as is the involvement of all levels of government and a wide range of non-government organizations" (Commonwealth of Australia 2007) (see Box text).

Moreover, recent advise to the Dutch government (VROM-raad 2007, after WRR 2006) assigns the spatial environment as the domain in which to deal with climate change as a wicked problem.

Both the facts that spatial systems are complex adaptive ones and climate change is seen as a wicked problem are in current spatial planning practice not genuinely recognised and implemented yet. However, in dealing with wicked problems thinking in complex adaptive systems offers the opportunity to escape the fixed, end-solution based practice to date (Roberts 2000). This is the reason to focus in this book on the possible connections between wicked problems (e.g. climate change) and spatial planning and design.

The Commonwealth of Australia (2007) states: "Tackling wicked problems is an evolving art. They require thinking that is capable of grasping the big picture, including the interrelationships among the full range of causal factors underlying them. They often require broader, more collaborative and innovative approaches. This may result in the occasional failure or need for policy change or adjustment". Tackling wicked problems is an evolving art but one which seems to at least require:

- Holistic, not partial or linear thinking.
- Innovative and flexible approaches.
- The ability to work across agency boundaries.
- Increasing understanding and stimulating a debate on the application of the accountability framework

Box Text – From Tackling Wicked Problems, Commonwealth of Australia 2007

Climate Change—A Wicked Problem
One issue that illustrates many of the characteristics of wicked problems is the current debate about the causes of and solutions to climate change. The debate has been simplified into three competing 'stories', which emphasise different aspects of the climate change issue.[1] Each 'story' tends to define itself in contradistinction to the other two policy stories and proposes different policy solutions.

- **Profligacy.** This is the story that sees prevailing structural inequalities, particularly between countries, as having led to increasingly unsustainable patterns of consumption and production. In this story, urgent fundamental reform of political institutions and unsustainable lifestyles is required. Decision-making needs to be decentralized down to the grass roots level and citizens need to dramatically simplify their lifestyles to conserve the earth's resources. The onus is on advanced capitalist states to take action.
- **Lack of global planning.** This story sees as the underlying problem the lack of global governance and planning that would rein in global markets and factor into prices the costs to the environment. It makes no sense for any household, firm or country to unilaterally reduce its emissions, as each individual contribution is too small to make a difference. Remedying climate change would require all governments and parliaments to formally agree on the extent to which future emissions should be cut, and how and when. States would then impose these formal intergovernmental agreements on the multitude of undiscerning consumers and producers within their borders.
- **Much ado about nothing.** This story sees much of the debate as scaremongering by naïve idealists who erroneously believe the world can be made a better place (profligacy story), or by international bureaucrats looking to expand their budgets and influence (lack of global planning). Some with this view are sceptical about the diagnosis of climate change itself, while others are convinced that, even if correct, the consequences will be neither catastrophic nor uniformly negative. Technological progress, adaptation and dynamic markets are the solution to the negative effects of climate change.

The three stories tell plausible but conflicting tales of climate change. None of the stories are completely wrong, yet at the same time none are completely

---

[1] The three different stories of climate change are identified.

(continued)

> **Box Text – From** (continued)
>
> right—each story focuses on some partial aspect of the debate. The stories' proponents are unlikely to agree on the fundamental causes of and solutions to the global climate change issue. And since these stories contain normative beliefs (either in egalitarian structures, in hierarchical bureaucracies, or in markets) they tend to be immune to enlightenment by scientific facts. This leaves the policy maker with a dynamic, plural and argumentative system of policy definition—typical of many wicked policy problems.

- Effectively engaging stakeholders and citizens in understanding the problem and in identifying possible solutions.
- A better understanding of behavioral change by policy makers.
- A comprehensive focus and/or strategy.
- Tolerating uncertainty and accepting the need for a long-term focus.

Many of these capabilities are often absent in regular spatial planning practices.

1.6 Conclusion

This chapter illuminates the current practice of spatial planning and the results current processes lead to. The conclusion can be drawn that current planning processes lead to 'more of the same' urban design, urban regions and landscapes. Even if problems become complex and wicked, the spatial planning habits tend to copy solutions of the past.

The regular planning process consists of an analysis and accompanying program to plan for. This program is subsequently translated into a design. As long as the quantitative requirements are met, the plan meets demand and is judged as satisfying. New problems are fitted into this system. Even if the problems are wicked and cannot be described in quantitative terms they are made part of the quantitative design process (only) for the parts that can be translated. The rest of the problem is often denied or translated in prosaic words of poetry, sketching beautiful futures, but failing to make it to the hardware; the binding and directive parts of the plan. The wicked problem is torn apart and the most complex parts are set aside.

Figure 1.12 illustrates this process as a tame, straightforward way of planning. If wicked problems, such as climate change are seriously taken up in spatial planning, the room for flexibility and planning 'over time' needs to be increased. In the planning process a wicked bypass (Roggema 2008) is required to create this room.

As demonstrated in this chapter, spatial developments in Australia and the Netherlands suffer from current planning habits, leading to endless repetitions of urban developments. These newly realised areas fail to be prepared for the impacts of climate change because it is tackled as a tame, quantitative problem. The urban fabric becomes

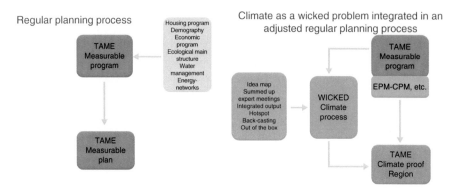

Fig. 1.12 Comparison of a tame planning process and a wicked one, in which is more room for turbulent adjustments (Roggema 2008)

a static and unchangeable pattern, in which the room for sudden floods or bushfires, prolonged droughts or heat stress is very limited. This, eventually, leads to disaster.

The choice of urban planners and landscape architects is therefore to start planning in a more dynamic way. In which the space for wicked problems is extended and spaces are created where the unexpected consequences of climate change can be mitigated. In order to create those spaces the urban fabric needs to be seen as a complex adaptive system, capable of self-organisation and change over time. The research how to develop and design those urban regions and landscapes is ongoing, but it may be clear that landscape elements, infrastructure and public spaces need to be able to change and adjust themselves to new environments. This is a major challenge for spatial planners and designers.

References

Adams R (2009) Transforming Australian Cities for a more financially viable and sustainable future. City of Melbourne & State of Victoria, Melbourne

Alterra, KNMI, DHV (2008) Klimaatschetsboek Groningen en Drenthe. Klimaat voor Ruimte, Wageningen

Arthur WB, Durlauf SN, Lane D (1997) Introduction. In: Arthur WB, Durlauf SN, Lane D (eds) The economy as an evolving complex system II. Addison-Wesley, Reading, pp 1–14

Bureau Stroming BV (2006) Natuurlijke klimaatbuffers, adaptatie aan klimaatverandering, wetlands als waarborg. Vereniging Natuurmonumenten, Vogelbescherming Nederland, Staatsbosbeheer, ARK Natuurontwikkeling, Waddenvereniging

Commonwealth of Australia (2007) Tackling wicked problems; a public policy perspective. Australian Government/Australian Public Service Commission, Canberra

Conklin J (2001) Wicked problems and social complexity, Cognexus Institute, p 11. [online]: http://cognexus.org/wpf/wickedproblems.pdf. Accessed 13 Dec 2010

De Jonge JM (2009) Landscape architecture between politics and science, an integrative perspective on landscape planning and design in the network society. PhD thesis, Wageningen University/Uitgeverij Blauwdruk/Techne Press, Wageningen/Amsterdam

Department of Infrastructure (2002) Melbourne 2030: planning for sustainable growth. State of Victoria, Melbourne
DHV (2007) Ruimtelijke impact klimaatverandering, Toekomstbeelden voor 2050. Provincie Groningen, Groningen
Dodson J, Sipe N (2006) Shocking the suburbs: urban location, housing debt and oil vulnerability in the Australian City. Urban Research Program, Research paper 8. Griffith University, Brisbane
Dodson J, Sipe N (2008a) Unsettling suburbia: the new landscape of oil and mortgage. Vulnerability in Australian Cities. Urban Research Program, Research paper 17. Griffith University, Brisbane
Dodson J, Sipe N (2008b) Planned household risk: mortgage and oil vulnerability in Australian Cities. Aust Plan 45(1):38–45
DPCD (2008) Melbourne @ 5 million. Melbourne 2030: a planning update. State of Victoria, Melbourne
Gallagher R, Appenzeller T (1999) Beyond reductionism. Science 284:79
Gleeson B (2010) The greatest spoiler. Griffith Rev 29:57–66
Holland J (1995) Hidden order: how adaptation builds complexity. Addison-Wesley, Reading
Holling C (1992) Cross-scale morphology, geometry and dynamics of ecosystems. Ecol Monogr 62:477–502
Jørgensen SE, Straskraba M (2000) Ecosystems as cybernetic systems. In: Jørgensen SE, Müller F (eds) Handbook of ecosystem theories and management. Lewis Publishers, London, pp 249–264
KNMI (2006) Klimaat in de 21ste eeuw, vier scenario's voor Nederland. KNMI, De Bilt
Lenton T, van Oijen M (2002) Gaia as complex adaptive system. Phil Trans R Soc Lond B 357:683–695
Levin SA (1998) Ecosystems and the biosphere as complex adaptive systems. Ecosystems 1:431–436
Levin SA, Segel LA (1985) Pattern generation in space and aspect. SIAM Rev 27:45–67
Massoud Amin S, Horowitz BM (2007) Toward agile and resilient large-scale systems: adaptive robust national/international infrastructures. Keynote international conference on flexible systems management GLOGIFT-07: 'Flexibility with Business Excellence in the Knowledge Economy'. Noida, India
Mitchell Waldrop M (1992) Complexity, the emerging science at the edge of order and chaos. Simon & Schuster Paperbacks, New York/London/Toronto/Sydney
MNP (2005) Effecten van klimaatverandering in Nederland. MNP, Bilthoven
Moodie R, Whitney D, Wright M, McAfee A (2008) Melbourne 2030: audit expert group report. State of Victoria, Melbourne
Murray JD (1989) Mathematical biology. Springer, Heidelberg
O'Neill RV, DeAngelis DL, Waide JB, Allen TFH (1986) A hierarchical concept of ecosystems. Princeton University Press, Princeton (Monographs in population biology 23)
Presland G (2009) The place for a village; how nature has shaped the City of Melbourne. Museum Victoria Publishing, Melbourne
Provincie Groningen (2000) Provinciaal Omgevingsplan, vastgesteld op 14 December 2000. Provincie Groningen, Groningen
Provincie Groningen (2006) Provinciaal Omgevingsplan, POP 2 tekst en kaarten, vastgesteld op 5 Juli 2006. Provincie Groningen, Groningen
Provincie Groningen (2009) Provinciaal Omgevingsplan 2009–2013, vastgesteld op 17 Juni 2009. Provincie Groningen, Groningen
Rittel H, Webber M (1973) Dilemmas in a general theory of planning. Policy sciences, vol 4. Elsevier Scientific Publishing Company, Inc., Amsterdam, pp 155–169, 1973 [reprinted in Cross N (ed) Developments in design methodology. Wiley, Chichester, pp 135–144, 1984]
Roberts N (2000) Coping with wicked problems. Naval Postgraduate School, Monterey, California, Department of Strategic Management working paper

Roggema R (2007) Spatial impact of adaptation to climate change in Groningen; move with time. Province of Groningen, Groningen

Roggema R (2008) The use of spatial planning to increase the resilience for future turbulence in the spatial system of the Groningen region to deal with climate change. In: Proceedings of the 2008 UK systems society international conference – building resilience: responding to a turbulent world, Oxford, 1–3 Sept 2008

Roggema R (2010) 'Swarm Planning' – a new era in spatial design and planning for climate change. Key-note lecture DSE-climate adaptation seminar. The Royal Society of Victoria, Melbourne, 3 Dec 2010

Roggema R, van den Dobbelsteen A (2008) Swarm Planning: development of a new planning paradigm, which improves the capacity of regional spatial systems to adapt to climate change. In: Proceedings World Sustainable Building Conference (SB08), Melbourne

Russel JS (2011) The agile city: building well-being and wealth in an era of climate change. Island Press, Washington, DC

The Age (2010) Green land cut back as Melbourne grows much bigger, 29 July 2010. http://www.theage.com.au/victoria/green-land-cut-back-as-melbourne-grows-much-bigger-20100729-10wvi.html. Accessed 16 Dec 2010

The Age (2011) Melbourne's growth boundary under review again, 17 May 2011

VROM-raad (2007) De hype voorbij, klimaatverandering als structureel ruimtelijk vraagstuk. Advies 060. VROM-raad, Den Haag

Wheeler T (2010) Garden cities of tomorrow: upside down, inside out and back to front. Griffith Rev 29:5–13 <http://search.informit.com.au/documentSummary;dn=37612045324191;res=IELLCC> ISSN: 1839–2954

WRR (Wetenschappelijke Raad voor het Regeringsbeleid) (2006) Klimaatstrategie – tussen ambitie en realisme. Amsterdam University Press, Amsterdam

Chapter 2
Turbulence and Uncertainty

Rob Roggema

Contents

2.1	Introduction	26
2.2	Turbulence	26
	2.2.1 End Nineteenth – Beginning Twentieth Century	26
	2.2.2 Beginning Twentieth Century – Mid Twentieth Century	27
	2.2.3 Sixties and Seventies	29
	2.2.4 Eighties and Nineties	30
	2.2.5 Nineties and Early Twenty-First Century	31
2.3	What's Next, Beyond Turbulence?	32
	2.3.1 Internet – The Driver	32
2.4	Uncertainty	34
2.5	Moment Uncertainty	37
2.6	Conclusion	39
References		40

Abstract This chapter determines the current societal environment as turbulent. It illustrates that depending on the type of environment in a certain period, both urban patterns and the way spatial planning and design is practiced change accordingly.

This timeframe is characterised by the free exchange of products, goods, information and values between consumers, who are also producers. It brings society beyond turbulence and in an uncertain timeframe. Uncertainty, which is often treated with the search for more certainty though gaining more knowledge, is however not always reduced in this old-fashioned way. Uncertainty, and more specifically moment uncertainty can be better approached through increasing self-organisation of the system. Learning from swarms, capable of increasing resilience through collaborating in smart groups, can inform spatial planning in a way that the spatial

R. Roggema (✉)
The Swinburne Institute for Social Research, Swinburne University of Technology,
PO Box 218, Hawthorn, VIC 3122, Australia
e-mail: rob@cittaideale.eu

system, a complex adaptive system, also performs swarm behaviour and organises itself in smart collaborating groups of spatial elements.

Keywords Turbulence • Uncertainty • Spatial planning • Swarm • Self-organisation

2.1 Introduction

In many (semi-popular) publications the current timeframe is declared complex, uncertain and turbulent. Recent uprising in the Middle East, The Economic Crisis, Climate Change are all seen as elements that contribute to increased uncertainty and a turbulent world. In this chapter, at first, the concept of turbulence is introduced and used to describe the current timeframe. Secondly, uncertainty is described as the result of these circumstances and society, policymakers and politicians naturally react with an urge to increase the knowledge to regain certainty. However, it is argued that uncertainty is only temporary and this period can also be seen as an opportunity of freedom to change, restructure and redesign.

2.2 Turbulence

Emery and Trist have identified characteristics of organisational environments (Emery and Trist 1965). They distinguish four different 'causal textures': placid randomised, placid clustered, disturbed reactive and turbulent environments. Babüroglu (1988) added a fifth, the *vortical* environment. These textures are predominantly used to describe organisational environments. However, when the characteristics of these textures are linked with the qualities of determined urban design periods, they suddenly describe planning eras. This exercise has been executed by taking the Dutch planning history as object.

2.2.1 End Nineteenth – Beginning Twentieth Century[1]

In this period urban development and population growth went slowly. There existed a natural balance between an elite and the people. A small group of individuals took decisions. The majority of changes just happened and were dependent on coincidences: the right people on the right place. It was a placid, randomised environment: *"The simplest type of environmental texture is that in which goals and noxiants ('goods' and 'bads') are relatively unchanging in themselves and randomly distributed"* (Emery and Trist 1965). Ordinary people worked many hours each day and

---

[1] Periods based on: prof ir. S.J. van Embden in Stedebouw in Nederland, 1985.

they didn't have the time to worry about political decisions. After the industrialisation started (end of the nineteenth century) living standards rapidly worsened. Many people migrated into the cities to work there and lived in miserable situations. A central planning system did not exist and building was individually organised. When you needed a roof above your head, you just start building it wherever it suits you best. The result of this randomised development, small, compact settlements without water, energy or sewage, can be compared with the shadow cities (Neuwirth 2005), which are built nowadays in developing countries. The people who could afford it erected larger houses on the better spots or even second villa's outside the city. The boundary of the city, often walled, was a clear one. As the size of cities increased, diseases became big problems. This resulted in the need to start regulate. In this period the first examples of comprehensive planning were build. A good example is the Baronielaan (Fig. 2.1) in Breda, where an early project developer built the entire street in one time.

2.2.2 Beginning Twentieth Century – Mid Twentieth Century

In this period the balance between elite and the people stayed more or less the same. However, the need to plan and regulate increased due to growing urban problems, such as pollution and diseases. Early in this period the Netherlands adopted several new National Laws, such as the Housing law (1901) and the Education law (1904). The relations in society were clear: the elite made decisions, the majority of people stayed at work. The elite also enhanced the first urban designs. The architect Berlage, for example, became the city building master in the city of The Hague. He designed the plan for the entire The Hague area (Polano 1988). Still, the results of his work can be viewed all around the city. Part of these greater plans was the design and building of Garden Cities, such as Vreewijk in Rotterdam. In the 1950, after World War II the rebuilding of the Netherlands had to take place and this required a central planning system to house all the people. The quick and efficient building of attached houses in rows is typical for this time and can be viewed in nearly every village and city.

Spatial planning was executed in a top-down way. The government decided what was good and bad for the people and designed and realised the spatial plans. These plans were constructed, calculated and based on the best available techniques. The government, convinced to have the knowledge for the best solution, conducted these plans centrally. It gave room to all kind of visions and futurists. With a little fantasy the *'Grand Projets'* of president Mitterand can be placed in this planning tradition. The central design principle is to construct an end-image of the future on paper. The environment was still stable and clear but the efforts were not completely randomised anymore. They were focussed on those areas where the urban designs should be realised. Therefore, this period can be referred to as placid, clustered environment: *".... which can be characterized in terms of clustering: goals and noxiants are not randomly distributed but hang together in certain ways"* (Emery and Trist 1965).

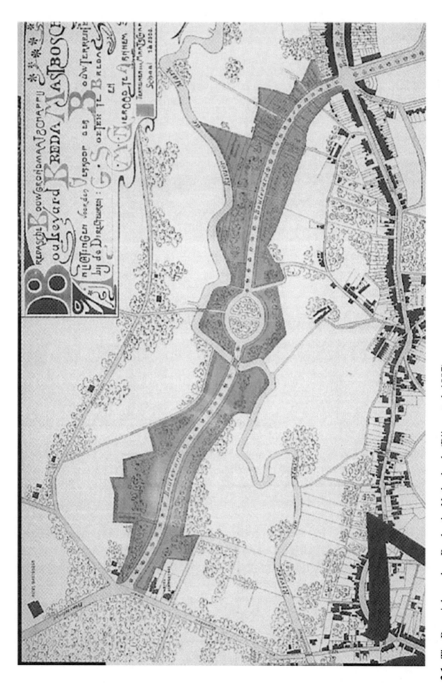

Fig. 2.1 The Baronielaan plan, Breda, the Netherlands (Bijma et al. 1997)

2.2.3 Sixties and Seventies

In this period a democratic transformation took place. Existing power balances in which the elite decided and the people had to obey was broken down, sometimes through radical protests, such as the *Maagdenhuis*-riots in Amsterdam or squatter-riots. Power became a shared resource obtainable for everybody and for all layers in society. This change had its effects on spatial planning as well.

The top down planning ultimately led to extensive areas with high-rise buildings, such as the *Bijlmer*-area in Amsterdam-southeast (Fig. 2.2) and *Nieuw-West* in Amsterdam-west. These areas, with spacious houses and modern transport systems were successful at first, but at later stages became the ghettos of Amsterdam, where minority groups, poor people and illegal refugees lived.

Because of the dissatisfaction with the results, the belief took root that the one and only top down solutions were not always the ultimate design answers. Government, in collaboration with enterprises, started to use scenarios, because the future is difficult to predict and planning in many occasions fixed end-images did not satisfy. Scenario planning showed several possible futures. Every scenario told a logical and consistent story. Still, the scenarios were constructed on the basis of technical and quantifiable information. Future, social or environmental demands were excluded. In practice, the choice of a certain scenario determined the way the old familiar, centrally directed, planning machine could continue its work. Despite the increase of complexity of the environment, planning practice continued a governmental concern. Turbulence occurred outside the planners' offices. In Emery and Trists words a disturbed-reactive environment: *"it is an environment in which there is more than one organisation of the same kind; …. The existence of a number of similar organisations now becomes the dominant characteristic"* (Emery and Trist 1965).

Fig. 2.2 Bijlmermeer area, Amsterdam-southeast (Picture: © Rob Roggema)

2.2.4 Eighties and Nineties

In this period 'democratisation' of the planning process took place. Power became divided, not without fight, over many layers in society. Urban planning became the subject of the fight between vested interests and those of all. The violent protests against the new urban plan for the *Nieuwmarkt* in Amsterdam highlighted exactly this tension. Everyone who had a certain interest, or not even that, could join deliberations about the future. In this phase 'consultation of the public' was formalised in laws, resulting in endlessly talking parties about the colour of the nearest flowerpot. Many meetings had to be held to reach a common decision. Politicians became talking machines, always looking for the compromise trying to please as many people as possible. To reach solutions they had to talk till they died, at least to late at night or the next morning.

Designs shaped the exact environment people demanded. The famous *Cauliflower*-neighbourhoods, such as the Haagse Beemden in Breda (Fig. 2.3), were planned and realised. Everyone in the Netherlands knows these areas: if you don't live there, you get lost. If you live in one of them you can find your way in every other

Fig. 2.3 Haagse Beemden: section of the lay out (De Boer and Lambert 1987) and typical view of a 'woonerf' (a car-free, dead-end street with street furniture, playgrounds and priority for pedestrians, cyclists and playing children) (Otten and Dijkstra 1989)

example in the country: as a result you might find yourself in someone else's house, thinking you are in your own. This bottom-up process of making everyone happy is more important than giving areas certain identities or designing an urban plan of a certain quality.

This environment consisted of many contrasting and interacting opinions, that all had to be taken seriously, but sometimes could not be united. There is no set of values determining good planning, which at least offers a common ground for deliberations. Therefore, in the words of Emery and Trist, this is a turbulent environment: *"the dynamic properties arise not simply from the interaction of the component organisations, but also from the ground itself. The 'ground' is in motion"* (Emery and Trist 1965).

2.2.5 Nineties and Early Twenty-First Century

In this period existing structures are unable to change, because they have the habit to reproduce themselves and perform repetitive patterns of working methods, which are well known and accepted both political as within governmental agencies. This fixed political system is no longer able to connect with the demands of today's society. The central government is confused. It understands that its citizens want to have a say in their urban environments, but they don't know how to allow their influence in planning. As a solution the government chooses to direct main developments such as the location where urban developments are allowed, determining the amount of houses and many other programmatic elements. Beyond the main decisions detailed decisions, such as the colour and architecture of the building (see Fig. 1.7, Chap. 1), are left to local authorities and citizens. It results in a basic quality of housing combined with a rich expression of urban patterns and identity-valued architecture. To execute these plans the government developed a new collaborative method: area development (Ruimtelijk Planbureau 2004), in which private developers, stakeholder groups and the government shared responsibility about the content, the investments and the realisation of a project.

Despite this effort to collaboratively share responsibilities, the environment becomes too complex. Obviously, not everyone can be included in the process. Only key players, people with strategic positions and other 'powerful' (but often unknown) men are collaborators, leaving the 'others' out of influence. Polarisation between the influential and the rest, objecting any development, is taking place. The ones with influence obey any unwritten rule that underpins their powerbase. There seems to be an underlying agreement, fixing existing power balance, but leading to repetitive processes, similar solutions and ongoing non-transparent decision-making. The 'rest', not involved, the process is non-transparent nor reachable. This is a *vortical* environment: *"… the prevalence of stalemate, polarisation and monothematic dogmatism … leads to a frozen or a clinched order of connectedness as well as that of unevenly dynamic turbulent conditions"* (Babüroglu 1988). It seems the elite-people dichotomy has reinvented itself.

In this situation the following problems occur:

1. The government still sets rules and is director, which leads to power orientated behaviour instead of goal-orientated behaviour;
2. It leads to fixed power constellations, not only inside a ruling political elite but also outside: a new conglomerate of 'important' people within a certain network: the new elite;
3. It focuses on the commonly defined and most easy-to-solve problems, not necessarily the main problems to be solved;
4. It does not make use of specific qualities of people hence it is not focussing on the unique contribution individuals may have.

2.3 What's Next, Beyond Turbulence?

Several developments determine the transition to a period beyond turbulence. The influence depletion of the natural resources, climate change and the Internet have, is causing not only turbulence, but feeds society with uncertainty and surprise. This environment can be identified as a novel. The energy system, climate change and the Internet-economy have similar characteristics. They:

- Are complex and difficult to overview and understand at once;
- Include lots of uncertainties;
- Are strongly interrelated with other functions and with each other;
- Have impact on the long-term.

Internet, Energy and Climate adaptation can be redefined using rules of Internet, where people are no longer seen as consumers, but as producers of information and value as well. What is valid in the Internet environment is true for producing energy or climate adaptation solutions: every citizen can be seen as producer of solutions and contributor of value.

2.3.1 Internet – The Driver

In the Internet-economy, people are no longer only consumers of news, adds or products, but they function also as a producer of information (Bakas 2005, 2006; NRC Next 2007; Eye Magazine 2007). Internet functions as a free space, where exchange can freely take place between consumers, who can change in a split second in producers and producers, who can change immediately into consumers. The society transforms from an industrial economy, based on the values of power, position and money, into an economy based on values and knowledge (Toffler and Toffler 2006; Greenfield 2003). Behind these economical defined changes different types of networks operate and cause changes in society. In the industrial timeframe the dominant network is centralised or, sometimes decentralised,

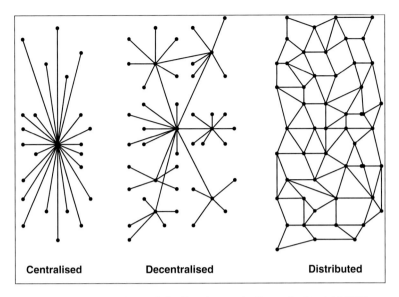

Fig. 2.4 Centralised, decentralised and distributed networks (Baran, in: Barabási 2003)

while the Internet-society is based on a distributed network typology (Fig. 2.4). As a result of these different network types, the impact individual people or collectives of individuals generate will have different outcomes. The, in current timeframe dominant, distributed network (trans)forms society into the *landscape 2.0* (Roggema 2008b).

The following transformations are identified:

- It is no longer useful to create an end-image of a society as it is in an economic sense no longer useful to think in end-products, because highest values are nowadays generated though inventive new assemblages of already existing parts, as has been proven in assembling the iPhone or iPad. The common idea rooted in many political movements that a certain end-view on how people should live and behave forms the basis for policy is also no longer fruitful. Instead, people need to be seduced to show certain behaviour. Society is more likely to emerge as a result of interactions between and the added behaviour of individuals;
- Climate change illustrates how an increasing series of complex interactions at a certain moment in time can lead to problems occur and only become apparent at a later stage. The exact relations between the interactions and the effects are difficult to overview;
- The new economy is a connection of people, ideas and information. In this new economy flexible network organisations become more important. In this context it is more important to be a connector of knowledge than an owner of goods. Possession is not the key factor. The key factors are the (mainly) immaterial additions to the network and the exchange of information;

- When the transition to an Internet-economy is used to understand future changes caused by climate change and energy supply, new landscapes lie in front of us. It is no longer only possible to consume landscape for living, enjoying or production, it can also deliver climatic resiliency and supply of sustainable energy to the spatial environment;
- People contribute individual pieces to a bigger world, knowing - partly unconsciously - that they are part of a bigger system, built out of billions of parts and consisting of unpredictable interactions. They start to understand that individual contributions and interactions form the future.

This change can be seen as a threat but it might also offer chances to adapt more easily to climate change, because individual people, under the condition that they are given the freedom to act, are able to contribute innovative solutions in dealing with the impacts of climate change and/or the production of sustainable energy. These new systems are no longer top-down, hierarchal and based on power, but are based on the value of everyone's contributions. This results in the emergence of a *'both ends'* society.

Changes in society have also implications for spatial planning practice. The unpredictability of interrelated long-term developments leads to a decreasing grip of the planning system on future developments, because traditional planning methods (blueprints, short-time oriented) are no longer useful (Roggema 2008a). What the world encounters is a society in which uncertainty plays a crucial role.

2.4 Uncertainty

Today's soccer community, like society in a broader sense, needs to operate in a turbulent and uncertain environment surrounded with uncertainties. Worries about contracting the best players, media income, government support and the individual choices of players where they want to play, all determine the dynamic context for soccer clubs. Different strategies are applied, ranging from expensive transfers of top players, search for undiscovered talents in less-known competitions or emphasising skill development of the clubs own youth. For many years, Ajax has been known and famous for its capability to develop, out of a large talent pool, international top-players. Its youth academy is seen as a high level nursery for talent and it still yields many talented players, but the results both in the National League as in European competitions underperform.

Therefore, the club asked its famous club-icon, Johan Cruijff, to analyse the problem and advise on the necessary changes to obtain new success. The reason for underperformance of the main squad, as he formulates it, was due to a hierarchical organisation in which the 'holy' Ajax-system had become the dogmatic standard, regardless the competitive demands or players skills. The clubs culture had become rigid, in which improvisation and anticipation were banned and a lack of innovation was apparent. In this context the trainers fulfil the qualifications of their appointment instead of train players to individually become better. The way out of this, according

Fig. 2.5 Ajax Dutch champions in 2010–2011 [http://richardvanhoek.com/item.php?itemId=325587]

to Cruijff, is to develop an organised chaos, in which self-organisation is the driving force. The aim must be to make each individual player better and skill each of them according the old Ajax standards of a surplus of technique, creativity and understanding. To reach this goal he introduces a *'phantom-team'* of collaborating trainers and staff around each individual, aiming to improve the player. Procedures and rules are put in place only to support this learning process.

Despite the fact that his advise cannot be directly linked to performance on the field, 2 months after the release of his report (Cruijff 2011), Ajax became Dutch champions for the first time in 7 years (Fig. 2.5).

Paraphrasing the Ajax analysis, spatial planning and climate change science operate in a complex and uncertain environment. This uncertain environment causes fear, beautifully articulated by Arthur Docters van Leeuwen (Sengers 2005): the political elite is afraid. The general response to this, and this can be seen throughout the spatial planning as well as the climate change community, is to try to reduce uncertainty by increasing procedures, developing more detailed models and controlling processes. However, gaining more detailed knowledge does not always increase certainty, or as Kevin Trenberth (2010) puts it: 'More knowledge less certainty'.

Climate change and climate adaptation are often linked with uncertainty. WG I of the IPCC, cited by the Global Commons Institute (2011), states: "Climate change, the greatest threat to mankind, is resistant to reliable methodological quantification. In many cases it is not possible to "ascertain the probability of outcomes and their consequences through well-established theories with reliable and complete data". Both the risk and uncertainty of climate change require a very large degree of subjective judgement, erring on the side of precaution". People are generally averse to uncertainty and vagueness and are accordingly reluctant to take action in response. However, if uncertainty is framed positively, people have stronger intentions to act (Morton et al. 2011). Different types of uncertainty are distinguished (Solomon et al. 2007; Dessai and Van der Sluijs 2007):

1. Value uncertainties – Is defined by the IPCC as the confidence in scientific understanding (Solomon et al. 2007). Value uncertainties (or statistical uncertainties (Dessai and Van der Sluijs 2007)) are generally estimated using statistical techniques and are expressed probabilistically. To cope with this type of uncertainty forecasting is often used (Engau and Hoffmann 2011).
2. Structural uncertainties - Are generally described by giving the authors' collective judgment of their confidence in the correctness of a result (Solomon et al. 2007). However, Jones demonstrates that if a high level of uncertainty precludes quantification, such as in case of rapid climate change or climate surprises, a probability cannot be attached, but despite the fact that timing and degree of these events are unpredictable, they may be planned for as they occur within a broad statistical framework (Jones 2000). This type of uncertainty is often dealt with using scenarios, reason why Dessai and Van der Sluijs (2007) call this 'scenario uncertainty'. This type of uncertainty is dealt with addressing a constituent factor, increase influence and subsequently predictability (Engau and Hoffmann 2011).
3. Recognized ignorance – Are uncertainties of which we realize – some way or another – that they are present, but of which we cannot establish any useful estimate, e.g., due to limits to predictability and knowability or due to unknown processes (Dessai and Van der Sluijs 2007). According to IPCC (Solomon et al. 2007) this 'unpredictability' arises in systems that are either chaotic or not fully deterministic in nature and limits our ability to project all aspects of climate change. Or, as Garnaut (2008) puts it: "There is uncertainty when an event is of a kind that has no close precedents, or too few for a probability distribution of outcomes to be defined, or where an event is too far from understood events for related experience to be helpful in foreseeing possible outcomes". Examples of this type of uncertainty are accelerated sea level rise or the possible shut down of the thermo-haline ocean circulation. The strategy to cope with this type of uncertainty is to develop resilience and flexibility to endure the effects of unpredicted events (Engau and Hoffmann 2011).

The latter type of uncertainty offers the best opportunity to frame uncertainty in a positive way. The main reason is that development of resilience implies positive directed adjustments to current systems and thus implying opportunities to change. Taking this perspective in dealing with uncertainties we do not aim for uncertainty reduction, but try to deal with it in the best possible way. This can be easily connected with the 'school' of the resilience approach, which, in dealing with climate adaptation, accepts uncertainty and expects unanticipated surprises (Dessai and Van der Sluijs 2007). It also explains why there is a lack of attention to uncertainties in major adaptation research works (Adger et al. 2007, 2009) and is the reason why adaptation strategies can be effective even if (regional) climate predictions are not available (Dessai and Hulme 2004; Hulme and Dessai 2008; Dessai et al. 2009). Instead, the development of *robust* adaptation decisions (Dessai and Hulme 2007), or measures (Wilby and

Dessai 2010), is preferable and is functional, regardless the level or type of uncertainty. Robust decisions and measures are low regret and reversible, incorporate safety margins, employ 'soft' solutions, are flexible and mindful of actions being taken by others to either mitigate or adapt to climate change (Hallegatte 2009), Moreover, "the epistemological limits to climate and ESM (Earth System Models) predictions should not be interpreted as a limit to adaptation, despite the widespread belief that it is. Climate adaptation strategies can be developed in the face of deep uncertainties[2]" (Kabat 2008). This raises the question what priority should be given to reducing uncertainty as it may not be essential to manage climate change (Mearns 2010), both in a technical as science policy sense (Meyer 2011). In addition, it may even prove cost ineffective to wait for more precise knowledge. Pindyck (2006) demonstrates that if catastrophic impacts are included in cost-benefit models, early action favours over waiting to implement adaptation measures. This is especially the case if the point at which a catastrophic outcome occurs, is unknown.

2.5 Moment Uncertainty

Given the different types of uncertainty, the question remains how uncertain weather related events or natural disasters are. The severe hurricane Katrina and the eruption of the Grimsvatn volcano (Fig. 2.6), which both made headlines in 2011, were both surprising and caused a lot of trouble. But these types of disasters are in fact not uncertain, are well predicted and only the timing is a surprise.

For many natural disasters, but also for climate related impacts, the place and magnitude of the occurrence is very well known. Specific maps (Fig. 2.7) with the location and strengths of volcanoes and earthquakes as well as the location of

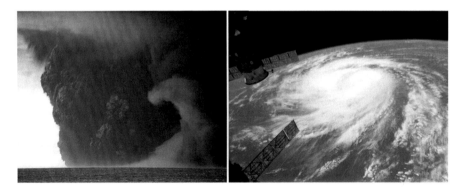

Fig. 2.6 Grimsvotn (May 2011) and Hurricane Katrina (September 2011)

---

[2] Deep uncertainty is defined as the condition where analysts do not know or the parties to a decision cannot agree upon (1) the appropriate models to describe interactions among a system's variables, (2) the probability distributions to represent uncertainty about key parameters in the models, or (3) how to value the desirability of alternative outcomes (Lempert et al. 2003, 2006).

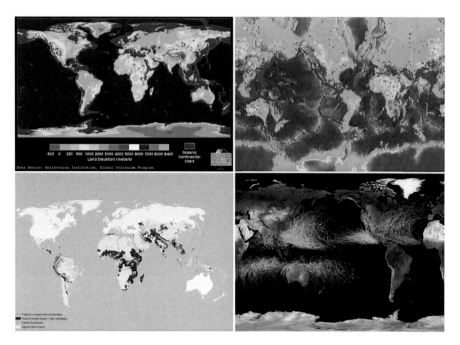

Fig. 2.7 Global distribution of Volcanoes, Earthquakes, Floods and Hurricanes

climate related events such as floods and hurricane trajectories explicitly point out the place where disasters can be expected. For the location of specific climate related impacts are detailed descriptions available for every centigrade extra global warming (Lynas 2007). The only element that is uncertain is the pace of climate change and the moment a disaster will happen.

Therefore, in addition to the three types of uncertainty, a new form of uncertainty needs to be taken into account: *moment* uncertainty, which can be defined as the uncertainty about the moment in time a certain event occurs. In dealing with moment uncertainty several strategies can be deployed. The first strategy is to accept that the event itself is well known, but when it will occur isn't. Similarly, the strategy to expect unanticipated surprises and the effects of unpredicted events will require anticipation and advanced preparation. The last strategy is to increase general resilience and flexibility in the projected affected area. When resilient designs are developed the area can withstand the moment the 'uncertain' event happens. Moreover, the resilience approach offers a positive framing of uncertainty. Highly resilient behaviour in nature may guide the development of the designs for our cities and landscapes. Swarms work together in smart groups and are capable of lessening the impact of uncertainty, complexity and change, e.g. increase their resilience (Miller 2010: 226). The objective here is to increase the resilience of the spatial system in order for it to deal with moment uncertainty. The resilience of the spatial system can be defined using the ecological definition of resilience: *"The capacity of a system to absorb disturbance and reorganise while undergoing change so as to still retain essentially the same function structure, identity and feedbacks"* (Walker et al. 2004).

In fact, this is exactly the same approach as proposed in the new organisational model for Ajax, where small phantom-groups work together and self-organise in order to increase the resilience of the player (e.g. his capability to deal at the pitch with sudden surprising situations, which he knows may occur, but during the match he only doesn't know when).

Learning from this model, spatial planning and climate change, and even more so if the two are linked, could profit from introducing self-organising principles. In order to be able to do so, the spatial system needs to be understood as a complex adaptive system, in which processes of self-organisation and emergence create ever changing spatial patterns, which, on their turn, can deal with unexpected change and uncertainty when circumstances ask for it, for instance as a result from climate change. 'Training' the individual spatial elements in the landscape with a surplus of 'technical skills' will create the spatial entities, which are capable of collaboratively increasing the adaptive capacity of the area through self-organisation and adapt more easily.

This type of response is a way of transforming to a new more complex mode of functioning, the most complicated way of reacting, as distinguished by Merry (1997). The other four are repeating former behaviour (1), varying behaviour slightly, predictably (2), adapting new behaviour (3) and transiting through a chaotic crisis (4). In order to design these responses spatially, possible far-futures need to be illustrated (Stremke et al. 2011), a long-term and abstract form of design. Transformational designs can only be translated in concrete short-term spatial interventions after having illustrated this far-future. Especially in dealing with moment uncertainty the far future is relevant, because the moment the region is confronted with an event can be soon or may be still a long time away. The character of illustrations, or abstract spatial designs, for a region anticipating moment uncertainty is to keep space literally open, for it to be used whenever the event occurs.

2.6 Conclusion

In a complex world and turbulent environments the methods and habits used in the past are no longer useful. As illustrated by the sequential urban planning periods in the Netherlands, every timeframe develops its own spatial planning approaches. The current timeframe, driven by the way on the Internet information and values is exchanged, is often defined as turbulent and complex. The response of many people, institutions and agencies is to strive for more certainty. A general (mis-)conception is that certainty is increased by collecting more knowledge.

In this chapter uncertainty is seen as a driver to search for new ways to plan and design. In the current timeframe there is no longer one ultimate solution for problems and a more flexible approach is required. An area needs to be seen as a combination of many functional elements, which in combination form a complex adaptive spatial system. The properties of those systems are that they can deal very well with turbulent environments and can produce responses that increase the resilience of the system. In dealing with a wicked problem, such as climate change, such as response is flexible over time and therefore capable of dealing with moment uncertainty, the uncertainty about the moment a climate event occurs.

The resilience of spatial systems (approached as complex systems) can be enhanced learning from the way swarms function. Swarms are capable of increasing resilience through collaboration in small, smart groups, hence lessening the impact of uncertainty, complexity and change. This is very similar to the way Johan Cruijff proposed to reorganise the coaching and training staff of Ajax in self-organising smart units.

Turbulence and uncertainty form two sides of the same medal. When the environment is turbulent it implies uncertainties. Instead of increasing the problem searching for more knowledge and certainty, the best approach is to navigate the problem. Whenever uncertainty appears, choose the best possible direction towards a far-future solution and keep on heading that way, whilst constantly adjusting the route.

Several aspects raised in this chapter are further explored in other chapters in this book. Complexity is further elaborated in Chap. 3, transformation in Chap. 4 and swarm planning Chap. 6.

References

Adger WN, Agrawala S, Mirza M (2007) Assessment of adaptation practices, options, constraints and capacity, Chap. 17. In: Parry ML, Canziani OF et al (eds) IPCC, 2007: impacts, adaptation and vulnerability, contribution of working group II to the fourth assessment report of the intergovernmental panel on climate change. Cambridge University Press, Cambridge
Adger N, Lorenzoni I, O'Brien K (eds) (2009) Adapting to climate change: thresholds, values, governance. Cambridge University Press, Cambridge
Babüroglu ON (1988) The vortical environment: the fifth in the Emery-Trist levels of organizational environments. Hum Relat 41(3):181–210
Bakas A (2005) Megatrends Nederland. Scriptum, Schiedam
Bakas A (2006) Megatrends Europe. Marshall Cavendish Business, Singapore
Barabási A-L (2003) Linked, how everything is connected to everything else and what it means for business, science, and everyday life. Plume, London
Bijma A, Hoeks D, van der Hoeven B, de Jong G, Joosse W, Lokerse H, Rensen M, Roggema R (Red) (1997) 100 jaar Baronielaan; Boulevard Breda Mastbosch. Sectie D, Zandbergse boekstichting, Breda
Cruijff J (2011) Op weg naar georganiseerde chaos, op weg naar succes. Rapport Technisch Platform, deel 1. Voetbalontwikkeling. Cruijff Inc., Amsterdam
De Boer H, Lambert D (1987) Woonwijken, Nederlandse Stedebouw 1945–1985. Uitgeverij 010, Rotterdam
Dessai S, Hulme M (2004) Does climate adaptation policy need probabilities? Clim Policy 4(2):107–128
Dessai S, Hulme M (2007) Assessing the robustness of adaptation decisions to climate change uncertainties: a case study on water resources management in the East of England. Glob Environ Change 17:59–72
Dessai S, van der Sluijs J (2007) Uncertainty and climate change adaptation – a scoping study. Copernicus Institute, Utrecht

Dessai S, Hulme M, Lempert R, Pielke R Jr (2009) Climate prediction: a limit to adaptation? In: Adger N, Lorenzoni I, O'Brien K (eds) Adapting to climate change: thresholds, values, governance. Cambridge University Press, Cambridge, pp 64–78

Emery FE, Trist EL (1965) The causal texture of organizational environments. Hum Relat 18:21–32

Engau C, Hoffmann VH (2011) Strategizing in an unpredictable climate: exploring corporate strategies to cope with regulatory uncertainty. Long Range Plan 44:42–63

Eye Magazine (2007) De nieuwe economie draait om liefde; Interview met Martijn Aslander. Eye Magazine 11–2007

Garnaut R (2008) The Garnaut climate change review: final report. Cambridge University Press, Cambridge/New York/Melbourne/Madrid/Cape Town/Singapore/São Paulo/Delhi

Global Commons Institute (2011) Climate uncertainty and policymaking. A policy maker's view. Report prepared for the All Party Parliamentary Group on Climate Change, London

Greenfield S (2003) Tomorrow's people: how 21st century technology is changing the way we think and feel. Penguin Books Ltd., London

Hallegatte S (2009) Strategies to adapt to an uncertain climate change. Glob Environ Change 19:240–247

Hulme M, Dessai S (2008) Ventures should not overstate their aims just to secure funding. Nature 453:959

Jones RN (2000) Managing uncertainty in climate change projections – issues for impact assessment, an editorial comment. Clim Change 45:403–419

Kabat P (2008) Should the uncertainty in climate scenarios limit adaptation? Presentation on 27 November 2008 during the congress 'On the road to a climate proof society', Rotterdam. http://promise.klimaatvoorruimte.nl/pro1/publications/show_publication.asp?documentid=3388&GUID=bb6eb9d4-1cec-4e48-9af3-d697abc1d213

Lempert RJ, Popper SW, Bankes SC (2003) Shaping the next one hundred years: new methods for quantitative, long-term policy analysis. RAND, Santa Monica

Lempert RJ, Groves DG, Popper SW, Bankes SC (2006) A general, analytic method for generating robust strategies and narrative scenarios. Manag Sci 52:514–528

Lynas M (2007) Six degrees, our future on a hotter planet. HarperCollins Publishers Ltd., London

Mearns LO (2010) The drama of uncertainty. Clim Change 100:77–85

Merry U (1997) Coping with uncertainty. Insights from new sciences of chaos, self-organisation and complexity. Praeger Publishers, Westport

Meyer R (2011) Uncertainty as a science policy problem. Clim Change. doi:10.1007/s10584-011-0050-8

Miller P (2010) The smart swarm. The Penguin Group, New York

Morton TA, Rabinovich A, Marshall D, Bretschneider P (2011) The future that may (or may not) come: how framing changes responses to uncertainty in climate change communications. Glob Environ Change 21:103–109

Neuwirth R (2005) Shadow cities. A billion squatters, a new urban world. Routledge, New York/London

NRC Next (2007) Werken zonder uurtje-factuurtje. NRC Next, 7 November 2007

Otten G, Dijkstra H (1989) Breda in panorama. Scriptum, Schiedam

Pindyck RS (2006) Uncertainty in environmental economics. NBER working paper series, Working paper 12752. National Bureau of Economic Research, Cambridge

Polano S (1988) Hendrik Petrus Belage, het complete werk. Alphen aan den Rijn, Atrium

Roggema R (2008a) Landschap 2.0. In: Roggema R (red) Tegenhouden of meebewegen, adaptatie aan klimaatverandering en de ruimte. WEKA uitgeverij B.V., Amsterdam

Roggema R (2008b) Swarm planning: a new design paradigm dealing with long-term problems, like climate change. In: Ramirez R, van der Heijden K, Selsky JW (eds) Business planning for turbulent times. Earthscan, London/Sterling

Ruimtelijk Planbureau (2004) Ontwikkelingsplanologie, lessen uit en voor de praktijk. NAi Uitgevers, Rotterdam

Sengers L (2005) De politieke elite is bang. Interview met Arthur Doctors van Leeuwen. Intermediair, 30 Juni 2005

Solomon S, Qin D, Manning M, Chen Z, Marquis M, Averyt KB, Tignor M, Miller HL (eds) (2007) Contribution of working group I to the fourth assessment report of the intergovernmental panel on climate change. Cambridge University Press, Cambridge/New York

Stremke S, Neven K, Boekel A (2011) Beyond uncertainties: how to envision long-term transformation of regions? In: Buhmann E, Ervin M, Mertens E, Tomlin CD, Pietsch M (eds) Proceedings digital landscape architecture 2011: teaching & learning with digital methods and tools. Wichmann Verlag, Offenbach\Berlin, pp 187–194

Toffler A, Toffler H (2006) Revolutionaire rijkdom, Hoe de nieuwe welvaart onze levens gaat veranderen. Uitgeverij Contact, Amsterdam/Antwerpen

Trenberth K (2010) More knowledge less certainty. Nature reports climate change 4 commentary http://www.nature.com/climate/2010/1002/full/climate.2010.06.html

Van Embden SJ (1985) Over vormgevers en vormgeving in de Nederlandse styedebouw van de 20ste eeuw. In: BNS (ed) Stedebouw in Nederland. De Walbrug Pers, Zutphen

Walker B, Holling CS, Carpenter SR, Kinzig A (2004) Resilience, adaptability and transformability in social-economic systems. Ecol Soc 9(2):5. [Online]: http://www.ecologyandsociety.org/vol9/iss2/art5/

Wilby RL, Dessai S (2010) Robust adaptation to climate change. Weather 65:180–185. doi:10.1002/wea.543

Chapter 3
Complexity Theory, Spatial Planning and Adaptation to Climate Change

Wim Timmermans, Francisco Ónega López, and Rob Roggema

Contents

3.1	Introduction	44
3.2	Complexity Theory	44
	3.2.1 Origins of Complexity Theory	44
	3.2.2 Concepts of Thought	45
3.3	Climate Change and Spatial Planning	48
3.4	Complex Adaptive Systems in a Spatial Planning Context	50
3.5	Four Planning Strategies for Climate Change	53
	3.5.1 Planning for Mitigation	54
	3.5.2 Sector-Based Adaptation	55
	3.5.3 Integrated Adaptation Planning	57
	3.5.4 Flexible Adaptation Planning, Swarm Planning	58
3.6	Discussion	61
3.7	Conclusion	61
References		62
Websites		65

Abstract This chapter analyses several concepts of Complexity Theory as regards their usefulness in spatial planning processes that aim to foster adaptation to climate change. The conditions under which a complex system is likely to change to higher

---

W. Timmermans (✉)
Groene Leefomgeving van Steden, Wageningen UR, Van Hall Larenstein –
Tuin en Landschapsinrichting, PO Box 9001, 6880 GB Velp, The Netherlands
e-mail: wim.timmermans@wur.nl

F.Ó. López
LaboraTe – Universidade de Santiago de Compostela,
Benigno Ledo s/n. – EPS, 27001 Lugo, Spain
e-mail: franciscojose.onega@usc.es

R. Roggema
The Swinburne Institute for Social Research, Swinburne University of Technology,
PO Box 218, Hawthorn, VIC 3122, Australia
e-mail: rob@cittaideale.eu

levels of complexity are seen as important when this system needs to deal with and adapt to climatic changes. This understanding is used to develop a framework in which these changes can be examined and explained. Supported by examples from various European countries, four different planning strategies (planning for mitigation, sectored adaptation, integrated adaptation and flexible adaptation) are positioned within the framework. We conclude that each of these strategies fills its own niche in the framework, that all strategies together describe the behaviour of a complex system and that flexible adaptation planning is most likely to facilitate a system change. When this reasoning is reversed and the question concerns which planning strategy fits in best with the demands imposed by climatic change (e.g. for a system change), flexible adaptation planning is seen as the most suitable option.

Keywords Complexity • Complex systems behaviour • Spatial planning • Climate change adaptation

3.1 Introduction

Climate change is seen as a growing threat to many of the world's cities in view of its associated negative impacts on their social and economic activities. Examples are the damage that sea level rise can cause to infrastructure, housing and drinking water provision; the increased impact of the urban heat island effect on public health; and the effect of extreme weather events and floods in many cities. Cities are not only the economic centres – more than half of the world's population lives in cities – but also the centres of decision and the creative centres of the world, so local authorities respond to these challenges with a wide range of approaches. This chapter discusses these approaches from the point of view of Complexity Theory. What are the shortcomings of these approaches when climate, cities and the adaptation of cities to climate change are considered as complex processes, and when current practices of related planning and design are taken into account? The chapter first discusses Complexity Theory and its implications for urban planning, then illustrates the current approach to climate change adaptation in cities, followed by a theoretical examination of design approaches. The chapter concludes by identifying knowledge gaps in design theory.

3.2 Complexity Theory

3.2.1 Origins of Complexity Theory

In 1956, Edward N. Lorenz, meteorologist and mathematician, was working on computerised simulations of the weather system. Once, when he wanted to continue a run that he had done earlier, he entered variables in the model that he had noted down near the end of the run of the day before. Surprisingly, the results, i.e. the "weather forecast", differed greatly from that produced the day before. When analysing the results, he discovered that the computer worked with six decimal places internally, but used only three in its presentation. This caused a rounding error of less than 0.001% in the start of the simulation, but led to an entirely different weather forecast (Lorenz 1963). This phenomenon became famous as the "Butterfly Effect": a butterfly flapping its wings in one place on earth could cause a hurricane elsewhere. A minute deviation in the initial variables caused by an unforeseen perturbation quickly leads to completely different weather conditions, which cannot be explained by linear thinking (Lorenz 1963).

Due to Lorenz' results, the prevailing paradigm, stating that everything in nature could be defined in exact figures, came under pressure. The rapid development of mathematics had meant that, since the seventeenth century, science had been based on the conviction that a given initial state of a system determines its entire future development. Natural phenomena were carefully observed and theoretically explained, after which these explanations were experimentally tested, accepted, adjusted or rejected. Lorenz was not the first one to challenge this paradigm. It had already been challenged by the findings of the French mathematician Poincaré (1854–1912), when he studied the so-called "Three-Bodies" problem. Newton had developed his mathematical laws, which describe the orbital paths of two heavenly bodies in relation to one another once their initial conditions are known. Following his line of thinking, Voltaire regarded the world as a mechanism, a timepiece. In contrast, Poincaré argued that it was mathematically impossible to predict the motion of three heavenly bodies in relation to one another on the basis of their initial conditions, shattering the euphoria about the world's predictability. From the 1980s, the "new science" of Complexity Theory gained attention in the academic literature. Research based on this theory encompasses multiple disciplines, sharing the notion of and focus on non-linearity and unpredictable events. Popular authors like Gleick (1987), Lewin (1992), Mitchell Waldrop (1983) and Cohen & Stewart (1994) have described the work of a wide range of scientists from different disciplines, who have discovered and researched non-linear phenomena and uncertainty.

3.2.2 Concepts of Thought

Complexity Theory describes and explains the behaviour of complex adaptive systems. In this chapter, Complexity Theory is used to improve our understanding of processes of spatial planning intended to foster adaptation to climate change. The chapter first describes the behaviour of complex adaptive systems, followed by a discussion of the use of Complexity Theory concepts in spatial planning and territorial development. This section ends with a detailed description of the characteristics of complex spatial planning and territorial development processes.

Complex adaptive systems are highly dynamic, and the interaction between the system and its context is vital for an understanding of its adaptive capacity. These systems can remain in a stable equilibrium for a long time – a state known as an attractor. Most literature on complexity concentrates on systems developing towards a higher level of complexity (e.g. Prigogine & Stengers 1984; Geldof 2001) while also devoting some attention to the possibility of decline. Systems change their structure slightly to adapt to external developments, in order to stay within their current attractor. While the system is in one attractor, there are other attractors (alternative states of form and operation) present to which the system could shift, but this only occurs after a shock that drives it out of its current attractor. Adaptation is a process internal to the system and is often described as self-organisation, happening within the system and potentially non-linear (Prigogine & Stengers 1984; Kauffman 1993). Self-organisation is the tendency of complex adaptive systems to evolve towards order instead of disorder; it only occurs in open systems that can import energy from external sources (Tiezzi 2003). Any complex system, although seemingly unchanged, is likely to become unstable as a consequence of changes in its environment. Instability does not equal change; it only means a growing likelihood that a shift will occur in some direction at a certain moment. When adaptation becomes increasingly difficult, the system develops into an unstable and chaotic state, which some complexity authors call "the edge of chaos". The system goes from one state of order (attractor) through this chaotic phase into another state of order or attractor (see: Mitchell Waldrop 1983; Geldof 2001). The change is rapid and chaotic, and its direction is unpredictable. This change is called a catastrophe (Scheffer et al. 2001) or a crisis (Geldof 2001).

One way to visualise the development of complex systems is illustrated in Fig. 3.1, representing the behaviour of a complex system over time. It shows equilibrium phases as well as sudden changes or crises. In the beginning (1), a complex system is in a particular equilibrium A. As a result of external factors (influx of energy or information), the system develops and reaches a less stable zone (2). The system tries to maintain equilibrium state A by suppressing change. At a certain moment, the system reaches a critical point where it turns into instability (3). Here, at the edge of chaos, the system looks for and finally moves to a new equilibrium, A' or B (4). It is uncertain what the new equilibrium will be.

The evolution of a complex system over time is schematically illustrated in Fig. 3.2. The system is in state $\times 1$. When this state becomes less stable or less favourable, the system quickly changes to a new state $\times 2$, with a higher degree of complexity. Another possibility is a development towards a lower degree of complexity, which is represented by the downward line. The shifts occur rapidly.

Prigogine and Stengers, who studied non-linear dynamic systems (Prigogine 1986; Prigogine & Stengers 1984) and Kauffman, who studied self-organisation (Kauffman 1993), are considered the founding fathers of Complexity Theory. The first researchers to adopt their ideas were natural scientists and physicists. As computers became more powerful, however, their ideas also found their way to researchers working on computer simulations of social systems and artificial life (Axtell and Epstein 1996). Nowadays, many scholars use features of Complexity Theory in studies of a broad range of natural and human systems. Complexity Theory is seen as a science enabling the gap between social and natural sciences

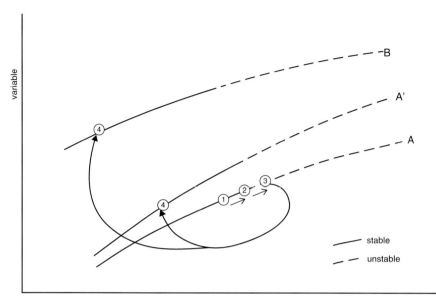

Fig. 3.1 Schematic behaviour of a complex open system with a certain characteristic (vertical axis) developing over time (horizontal axis). Tensions between the system's characteristics and its environment grow over time, changing the system from a stable to a more unstable state. The final result is a change in characteristics and a new stable balance with the environment (After: Geldof 2001; Prigogine 1986)

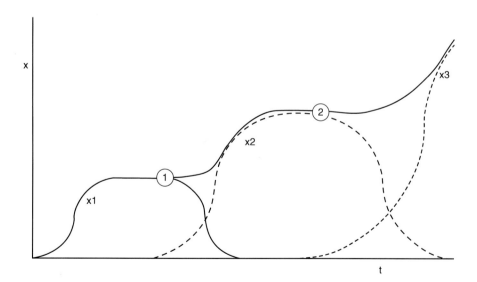

Fig. 3.2 Evolution of a complex system showing rising and declining complexity of the system (x) in relation to time (t) (After Geldof 2001; Prigogine 1986)

to be bridged (Lansing 2003; Liu et al. 2007; Nowotny 2005; Prigogine 1986; Urry 2006). Complexity Science has developed from studying closed systems to the study of open systems, including real-life situations, in a wide range of social sciences (Anderson 1999; Byrne 2003; Crawford et al. 2005; Duit & Galaz 2008; Levinthal & Warglien 1999; Montalvo 2006; O'Sullivan 2004; Plowman et al. 2007; Richards 2002; Teisman & Klijn 2008; Pulselli & Tiezzi 2009; Timmermans et al. 2011). The sudden unexpected change from one attractor into another, as well as aspects of dealing with the uncertainty of possible unexpected changes, have been thoroughly examined. Artigani (2005) discussed leadership, demonstrating how military leaders in history, operating on the basis of limited knowledge, were more successful when they were better prepared to adapt to unforeseen occurrences. Geraldi (2008) discussed the role of project management in making firms adaptive to discontinuous workflows and turbulent environments. Kelly & Stark (2002) examined numerous interviews to identify what factors make a company able to overcome a sudden extreme disruption in its environment, in their case the September 11 attacks; Anderson (1999); Vicenzi & Adkins (2000) and Mason (2007) discussed organisational vitality and leadership aspects in relation to complexity.

3.3 Climate Change and Spatial Planning

Climate change is widely seen as a long-term issue, having time horizons of several decades or even longer. Changes are expected to be associated with changes in the dimensions of hazards and disasters, which drive increasing social and economic losses, as people increasingly tend to settle in hazardous regions (Raschky 2008).

Urbanised regions are facing problems associated with their high population density and the concentration of economic activities and infrastructure. Coastal cities are susceptible to sea level rise and the risk of flooding – as seen in New Orleans or Bangkok – aggravated by the global processes. Mountainous regions can experience glacier retreat, with big impacts on water resources, as well as flash floods. Heat waves and droughts, and extreme weather events occurring with increased frequency and intensity, will also impact on cities. To date, very little knowledge of the social effects of climate change or catastrophic events is being considered at the city level, and most of the knowledge being gathered relates to cities in the developed world and concerns risks associated with sea level rise, water resources and health (Hunt and Watkiss 2011). An example of the latter is the very detailed quantitative risk analysis that is available for the metropolitan area of New York regarding sea level rise associated with climate change (Gornitz et al. 2002). However, most cities are applying sectored approaches (Ruth and Coelho 2007) and only a few cities have attempted to combine risk reductions into "city-level development strategies" (Hunt and Watkiss 2011). In general, the majority of adaptation plans suffer from a "narrow, climate-centric manner" (Preston et al. 2011). While Preston et al. (2011)

identified adaptation science as "inherently grounded in policy development and change", many authors have studied climate change adaptation on the basis of practical experiences at local and regional scales (Gornitz et al. 2002; Bedsworth & Hanak 2010; Hunt & Watkiss 2011; Nath & Behera 2011).

Rural areas are exposed to greater risks of wildfires and intense and prolonged droughts and floods. Moreover, people in these areas are more dependent on nature-based resources, which makes them highly vulnerable, due to a combination of multiple stresses, poor education and a lack of financial resources (Nath & Behera 2011). However, rural areas are beyond the scope of the present chapter, which concentrates on the metropolitan areas.

Mitigation and adaptation are two different strategies to deal with climate change. According to Füssel (2007), most attention to date has been devoted to mitigation measures, which aim to decrease the root cause of the problem of climate change and have effects on the vulnerability of all systems. Their success can easily be monitored, and the "polluter pays" principle can be easily applied. However, since the impacts of climate change are already manifest, the use of adaptation measures is inevitable. Moreover, changes (including climate change) in the Earth's system have always occurred and will always occur (Orlove 2005), making adaptation a historical necessity in any circumstance. Initially, adaptation approaches were developed in response to hazards or predicted risks, based on climate knowledge. The main result of these approaches has been awareness of the problem, resulting in a growing cooperation between scientists and practitioners (Füssel 2007). Various authors now state that successful urban or regional adaptation should involve adaptation measures which fit in and show synergy with strategic development approaches (Nath & Behera 2011).

The opportunities to invest in adaptation or mitigation measures are considered to be different for developed and developing countries and for urban and rural areas. Grimm et al. (2008) state that, in general, cities are in a special position: not only do they potentially suffer the largest negative impact, they are also causing the majority of the problems and they do have the potential to solve the problems in the long run, due to their concentration of economic activities and creativity.

Spatial planning is important for societies to adapt to climate change. In current planning practice, it is the local and municipal authorities that undertake most of the action for climate change adaptation (Füssel 2007; Ford et al. 2011), usually in their most sensitive sectors. As a consequence, many adaptation approaches have been based on a sectored point of view (Füssel 2007; Wheeler 2008; Ford et al. 2011). Despite the widely recognised necessity to develop more integrative approaches, taking climate change as "the local manifestation of a global problem" (Brace and Geoghegan 2010), most spatial planners in the UK are concentrating on flood risks. This means that they are not considering, and thus not integrating, wider implications (Wilson 2007), even though their thematic and non-integrated approaches are not very effective (Sanchez-Rodriguez 2009). Underdal (2010) identified three reasons why local authorities usually fail to take the global context into account: the long period between a particular human action and its environmental effect, the complex problem which is not yet fully understood, and the problem of scale which involves numerous small local actions impacting on the so-called global collective goods. The sectored approaches and the gap between local action and global impact mean that adaptation plans are greatly insufficiently developed (Preston et al. 2011). Spatial planning is widely considered to have to address this problem and to play an important role in achieving a proper balance between a "dominant agenda of economic growth and a less powerful discourse of environmental discourse" (Campbell 2006). Moreover, spatial planning is not only seen as responding to the demands for climate change adaptation, but can also offer a platform for the integration of "economic development, habitat protection and public safety" (Bedsworth & Hanak 2010), which are problems that greatly interfere with the problem of climate change (Orlove 2005). As a result, there is a growing interest in linking adaptation to long-term integrated development plans, either social or spatial (Nath and Behera 2011; Sanchez-Rodriguez 2009). The importance of "the ability of societies to adapt is determined … by the ability to act collectively" (Adger 2003), and spatial planning establishes the framework within which this collective awareness and action can take place.

Local action to address climate change adaptation is necessary and has to be integrated in long-term development plans (Sanchez Rodriguez 2009), taking into account the problems of long-term governance (Underdal 2010). Climate change adaptation requires a type of spatial planning which is oriented on long-term developments, devoting a great deal of attention to long-term governance and also focusing on tackling economic, demographic and societal problems. This implies that spatial planning for climate change adaptation requires a complex adaptive systems approach. In this chapter, the behaviour of complex adaptive systems is used as the framework within which planning for mitigation and adaptation policies can be understood as complex planning processes.

3.4 Complex Adaptive Systems in a Spatial Planning Context

Traditionally, spatial planning was considered as a technical discipline, in which people and space were represented as value-free and objective (Hillier 2008). From the 1970s onwards, spatial planning evolved towards more communicative approaches. Stakeholders became increasingly important in the planning process, resulting in a larger role for debate and dialogue. As a consequence, strategic and scenario planning, which were capable of dealing with the growing sense of uncertainty, became an important part of most spatial planning processes. Instead of value-free facts, communication and power became highly decisive for the outcome of the planning process (Faludi 2004; Healy 1997, 2003; Innes 2004). As a result, both theorists and practitioners of spatial planning have become aware of the fact that, although spatial planning is deeply rooted in a control paradigm, the outcome of a planning process can differ greatly from the intended outcome. Hence, the results can be highly surprising and spatial planning intrinsically has to deal with uncertainty and fuzziness (Timmermans 2004; De Roo & Porter 2007; Roggema & Van den Dobbelsteen 2008; De Jonge 2009). This chapter explores complex planning processes, that is, planning processes dealing with uncertainty and fuzziness, and tries to understand them in terms of complex adaptive systems.

Spatial planning processes are complex, as they need to deal with numerous levels of scale and subsystems. These processes need to be seen as constituting an open system, which continuously adapts to frequently interfering external forces that try to influence them. Spatial planning has to deal with natural and human systems, which interact everywhere and at any moment. Physical and chemical processes are apparent everywhere on Earth, while biological processes can be regarded as life and social processes as complex life, and intellectual processes are considered as specific to humans (Geldof 2001). Going from physical to intellectual processes implies rising complexity, so spatial planning and territorial development are highly complex systems. A rich literature is available dealing with the complex (and uncertain) features of spatial planning (Crawford et al. 2005; O'Sullivan 2004; O'Sullivan et al. 2006; Portugali 2006; Manson & O'Sullivan 2006).

How can we understand a planning process in terms of complex adaptive systems? Timmermans and others have conducted several research projects in the Netherlands and Galicia studying unexpected results of initially traditional planning processes (Timmermans 2004; Woestenburg et al. 2006; Timmermans 2009; Timmermans et al. 2011). The studies were based on informal in-depth interviews with key persons in the planning processes, a method described in Grounded Theory (Glaser & Strauss 1967; Strauss & Gorbin 1998; Charmaz 2006; O'Connor et al. 2008) and used to gain more insight into the informal planning process being studied. As mentioned above, planning is deeply rooted in control. As a consequence, planners with hindsight tend to construct a formal linear history of the planning process, omitting numerous non-linear events.

The example of the Dutch town of Breda (Timmermans 2009) was used as a case study to draw up a schematic outline of the behaviour of complex planning systems, using open in-depth interviews to construct a narrative of the planning process. To the west of Breda, between the municipalities of Breda and Prinsenbeek, plans were made to upgrade the highway from four to six lanes. A major part of the plan was a high flyover near housing areas. In spite of considerable social unrest, the plans were ready for implementation in 1996. At national level, however, a decision was suddenly taken to develop a new high-speed railway line connecting Amsterdam with Paris, which was to cross the same area of Breda. One Ministry was responsible for both planning processes, although two different departments were in charge of each of them. At the start of the process, attempts were made to ensure that the two plans would not interfere with each other, which resulted in an additional and even higher flyover near the housing area. Despite all efforts to prevent disruption, the two planning processes started to interfere in such a way that the implementation of the highway works was delayed. Public protest in Breda rose again, which resulted in a decision by the Breda municipal authorities that they would no longer cooperate with the new high-speed rail plans. This attitude changed, however, when municipal boundaries were redrawn, which meant that Breda and Prinsenbeek, situated on different sides of the highway, became one municipality. The large new infrastructure now became a threat, because it divided the new municipality into two parts. At this stage, cooperation with the Ministry was re-started. After a massive protest in Prinsenbeek village, the national government offered additional funds to allow better physical integration of the high-speed rail line into the area. A brainstorming process started, involving more than a hundred policy makers, and resulting in the proposition to create a new station for a shuttle train to connect Breda Central Station to the high-speed rail line, and two large green urban park corridors crossing the highway and the railway line, to connect Breda and Prinsenbeek.

This narrative can be rephrased in terms of complex open systems behaviour. The initial high-speed railway line planning process was a current routine (1) and in terms of Complexity Theory it was the initial, existing attractor. The project leader faced serious pressure (2) at the municipal level, as his project started to interfere with another complex project. The result was that the overall planning process became increasingly complex. He initially tried to adapt his project to the external pressures within the current routine (3). However, the external pressure forced the initial planning process into a chaotic phase (4). The interference between the two planning processes, the redrawing of the municipal boundaries and the public protest which resulted in extra funding being made available, became the triggers (5) for a sudden shift. These triggers caused the initial project to suddenly shift in an unplanned and rapid way (6) to another, previously unknown attractor (7) with a higher level of complexity. The outcome of this process was limited to a high-speed railway line, as had initially been planned. The planning process behaved like a complex adaptive system and developed according to Complexity Theory patterns, as illustrated in Fig. 3.3. An innovative wildlife bridge, the "Breda urban green corridor", unexpectedly became a logical additional result of this planning process.

Figure 3.3 presents the framework for typical complex systems behaviour in a spatial planning context, using the characteristics of complex adaptive systems developing over time.

This typical behaviour of a complex system in a spatial planning context consists of the following phases:

- The current routine of the complex system (e.g. current attractor, 1).
- Changes in the environment of the system resulting in pressure to change its routine (2a, 2b, 2c).
- Micro-scale attempts within the complex system to adapt to the outside changes while remaining within the current routine (3).
- Chaotic phase of increasing pressure in the system, in which current routines are no longer functional (4).
- Triggers from outside or inside of the complex system, sudden occurrences enhancing change (5).
- Sudden and rapid changes to the routines, evolving in a dominant direction towards a preferred, but uncertain and unknown new routine (6).
- The new routine of the complex system (7, which is the new 1).

3 Complexity Theory, Spatial Planning...

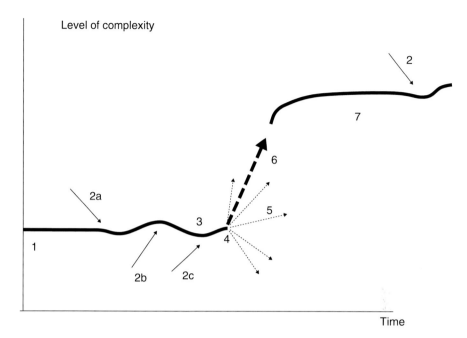

Fig. 3.3 Typical behaviour of a complex system, changing under outside or peripheral pressure from one attractor to another via a crisis (Timmermans 2009)

3.5 Four Planning Strategies for Climate Change

Section 3.3 discussed mitigation, sector-based adaptation and integrated adaptation in relation to spatial planning. Section 3.4 briefly discussed spatial planning approaches and considered spatial planning processes as complex processes dealing with uncertainty. The present section considers mitigation, sector-based adaptation and integrated adaptation as planning strategies in terms of complex planning processes, and illustrates them. It also adds and discusses a fourth adaptation planning strategy, namely flexible adaptation planning or swarm planning, which we consider necessary to deal with uncertainty and new and unforeseeable events. In the light of current practices in cities which are developing strategies to adapt to climate change, the question is how planning processes, understood as complex dynamic systems, can help to evaluate current practices and develop future climate planning strategies. Each of the four strategies, mitigation planning, thematic adaptation planning, integrated adaptation planning and adaptation planning in an uncertain environment, can be positioned in the framework of complex systems behaviour in a spatial planning context. These strategies are illustrated below by means of specific planning examples, three of which (Vouga, De Wijers and Varna) are derived from

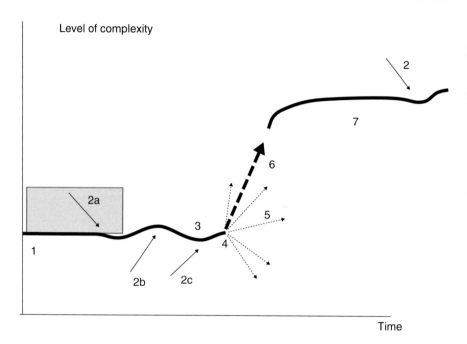

Fig. 3.4 Planning for mitigation positioned within the framework for complex systems behaviour in the context of spatial planning

the F:ACTS! project [www.factsproject.eu] and one, the "Floodable Landscape", from regional planning practice in the Netherlands. The strategies and examples are presented in the next sections.

3.5.1 Planning for Mitigation

Mitigation measures seek to take away a cause of climate change and its effects; an example of mitigation is the underground storage of CO_2, which is one of the major causes of anthropogenic climate change. Although mitigation measures can have some impact on territorial planning processes, their influence on the planning process is not very large. Since mitigation is beyond the scope of this thesis, we only briefly consider this strategy.

How can we interpret planning for mitigation in terms of complex systems behaviour (Fig. 3.4)? Planning for mitigation mainly focuses on the climate change issue (2a), which is represented in the framework as one of the external pressures. This planning strategy aims to minimise the effects of climate change and to keep the system within its original attractor (1). Mitigation planning does not necessarily have a territorial component.

An example of planning for mitigation is found at *De Wijers*, Belgium. This intimate rural landscape in the highly urbanised Hasselt–Genk area faces severe impacts of climate change [http://www.factsproject.eu/pilotprojects/dewijers/Pages/default.aspx], especially floods (Fig. 3.4) and droughts. One of the main objectives of the master plan which has been developed for the area is to make the entire province carbon-neutral by 2020. This ambition is to be realised through a strong emphasis on renewable energy supply, such as the use of biofuels.

In general, it is national governments or higher-level organisations which adopt mitigation measures, as the benefits of investments lie at the global level. In contrast to this general pattern, the regional authorities in *De Wijers* proposed mitigation measures not only for local benefits but to set an example and to encourage higher-level authorities to follow it. In addition, the proposals were seen as a message to individual citizens, to create awareness.

3.5.2 Sector-Based Adaptation

Despite mitigation efforts, societies will face the impacts of climate change in the future and therefore need to adapt. Initially, adaptation efforts were sector- or discipline-driven. These sector-based adaptation plans often try to confront an immediate or short-term climate change related problem (Sanchez-Rodriguez 2009). The usual process by which a city operates is that it first defines its vulnerability to climate change and assesses the related risks, then identifies adaptation options, carries out a cost-benefit analysis and finally develops a climate change adaptation strategy, within which a package of measures is proposed. After the strategy has been adopted, it is implemented, monitored and evaluated. The consequence of this approach is that the majority of climate change adaptation measures are stand-alone and expensive. These measures often deal with a single climate problem, such as reducing the "urban heat island" effect, preventing flooding due to severe storm events or tackling the effects of sea level rise.

As Fig. 3.5 shows, sector-based adaptation planning can also be interpreted in the context of complex systems behaviour. Adaptation is strongly related to territorial planning. Sector-based adaptation planning is mainly limited to climate-related technical measures (2a). Like mitigation planning, this planning strategy aims to minimise negative effects of climate change and to keep the system within its original attractor (1). The sector-based adaptation measures are included in complex planning processes as relatively technical actions in the stable linear phase, which is why this planning strategy is placed in the stable phase of the framework, although some proposed measures are more innovative and move the system to a less stable phase (3).

The Vouga region in Portugal is characterised by a so-called "bocage" landscape, traditionally based on small-scale and diverse forms of farming (Fig. 3.6) [http://www.factsproject.eu/pilotprojects/baixovougalagunar/Pages/default.aspx]. The marshland in this region is a valuable ecosystem which is unique in Portugal.

Fig. 3.5 De Wijers area

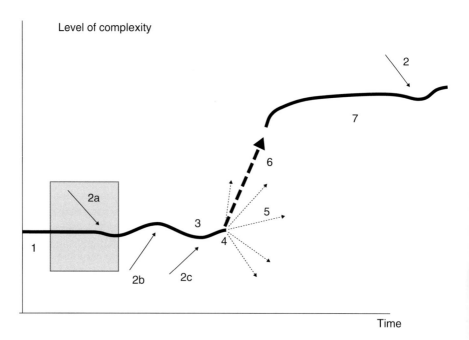

Fig. 3.6 Planning for sector-based adaptation positioned within the framework for complex systems behaviour in the context of spatial planning

Fig. 3.7 Vouga area

The wetlands are highly dependent on a critical balance between marine salt water and fresh water from the Vouga river. The effects of climate change, such as sea level rise, may cause saltwater intrusion and disturb this vulnerable balance, while extreme weather events, in combination with increasing urbanisation, cause severe river floods. These immediate sector-based problems are addressed by means of sector-oriented measures, involving floodwater protection measures and a dike to prevent salt intrusion.

3.5.3 Integrated Adaptation Planning

Recent adaptation planning has come to be regarded by the pioneers of urban climate change adaptation as an integrated part of urban development strategies and policies (Sanchez Rodriguez 2009). Climate change adaptation strategies are integrated within strategic, socio-economic and spatial development policies. These strategies contribute to and match climate change adaptation options and the city's strategic agenda. In this integrated planning approach, existing plans such as those for city parks, urban green spaces and tree-lined avenues offer additional opportunities for climate change adaptation goals, such as addressing the urban heat island effect. Strategic actions (and investments) are provided with adaptation goals.

As Fig. 3.7 shows, integrated adaptation planning can be interpreted in terms of complex systems behaviour. Integrated adaptation can be positioned as a part of traditional long-term strategic planning aiming at a stable and largely linear development (1). However, various external pressures such as climate change and changes in demography and socio-economics (2a, 2b, 2c) mean that complexity is increasing, with growing numbers of powerful stakeholders becoming involved, who try to interfere in the planning process. The consequence is by definition a less stable and non-linear communicative planning process (3). The joint planning efforts share the major goal of preventing the system from reaching a tipping point (4).

For example, the effects of climate change in the Varna region (Bulgaria) were investigated and brought to the attention of stakeholders with common interests. The main objective was for the new General Spatial and Land Use Plan for Varna to include a regional water-saving policy and the provision of water of sufficient quality throughout the summer months. These aims are no longer seen as a sector-based necessity, but as a part of the economic viability and branding of Varna as the main "Black Sea coast cultural and touristic bathing region". The integrated adaptation planning in Varna is "an example of a climate adaptive planning practice for a peri-urban area with different stakeholders and different levels of authorities involved", in which land use planning, drinking water, water management, sewage treatment and tourism are linked.

3.5.4 Flexible Adaptation Planning, Swarm Planning

The planning strategies described above are all rooted in the traditional planning paradigm of control and attempting to exclude instability and non-linearity as much as possible. However, climate change is considered to be a "wicked" problem (VROM-raad 2007; Commonwealth of Australia 2007), or even "super-wicked" (Lazarus 2009). There is still uncertainty about the rate of climate change to be expected, which implies that a broad variety of impacts can potentially occur (Lynas 2008). As a consequence, the level of uncertainty that spatial planning has to deal with is high. Moreover, interference from other problems, such as demographic processes, the economy or other aspects, implies an even greater uncertainty in the long term. This may have serious consequences for the stability of long-term planning processes.

As Fig. 3.8 shows, flexible adaptation planning can be interpreted in the context of complex systems behaviour. If complexity keeps growing, and uncertainty increases, planning may become – or may need to become – unstable. Only then will it be able to respond to uncertain circumstances or uncertain future expectations. The goal of planning is to anticipate various expected crises (4). An "out of the blue" mode of thinking can be used to predict extreme events and find possible ways to handle them (5). As a consequence, changes can be planned to happen slowly and in a controlled fashion (6a) instead of being extreme and unexpected (6b). This can involve a previously "unthinkable" measure, such as allowing the future climate event (e.g. a flood) to happen now (a floodable landscape). It implies that the planning process needs to be changed completely. This can happen as a consequence of climate change or of insights into the dramatic long-term consequences of climate change, or as a consequence of any other type of crisis. Planners increasingly work with long-term scenarios which include unexpected developments.

3 Complexity Theory, Spatial Planning... 59

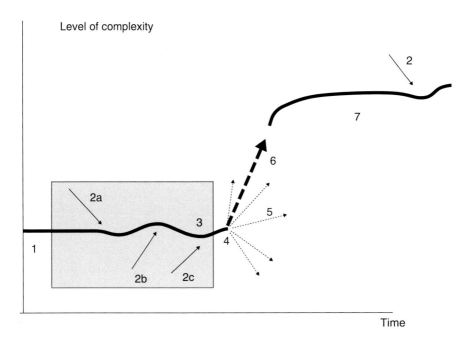

Fig. 3.8 Planning for integrated adaptation positioned within the framework for complex systems behaviour in the context of spatial planning

The proposals for a "Floodable Landscape" (Roggema 2008, 2009a, 2009b, 2010, 2011; Roggema et al. 2011), an example of swarm planning, take this uncertainty as the starting point for adaptation planning. The Eemsdelta (Fig. 3.9) is the weakest point in the coastal defence infrastructure in the northern part of the Netherlands, which means that if the sea level rises, this might be the first to be threatened by flooding in the case of a severe storm during spring tide. A traditional response to such a threat was to invest in heightening and strengthening the dikes, but even the strongest dike can be breached, and the stronger the dike, the worse the disaster. A different approach has therefore been proposed, in which instead of investing in higher and stronger dikes, small controlled floods would be allowed to occur through a deliberately made hole in the dike. Different areas would be made suitable for flooding for different levels of sea level rise. A long-term design and the ability to predict where the flooding will occur would thus enable adaptation at a very early stage. People, buildings and social organisation can be sufficiently prepared for small controlled floods. Instead of a reactive approach, which defends the coastline and waits for a disaster to happen, the proposal for a Floodable Landscape initiates action to anticipate climate change hazards. When an uncertain future can be anticipated in such a way that the landscape is prepared for whatever this future may bring, the uncertainty and potential climate change hazards can be outpaced. By allowing the future to happen, inhabitants, authorities and local groups will be able to adapt, even before the event occurs. The example of the Floodable Landscape is described in detail in Chap 8 (Fig. 3.10).

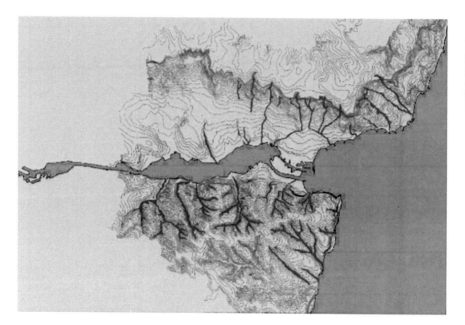

Fig. 3.9 Varna, floods

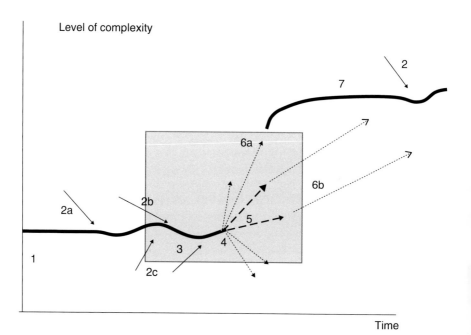

Fig. 3.10 Flexible adaptation planning positioned within the framework for complex systems behaviour in the context of spatial planning

3.6 Discussion

This chapter has described technical planning, communicative planning and planning with uncertainty. As regards climate change, planning for mitigation and sector-based adaptation focus on simple technical measures that have to be taken, such as carbon sequestration or flood mitigation. When the measures are widely supported, technical planning is appropriate, but when measures are not agreed upon in the public debate, communicative planning processes, with participatory stakeholders playing an important role, are the most suitable processes to use. Adaptation processes are more complex, mainly due to increased external influences from other relevant problems and increased participation of stakeholders in the planning process. Using a long-term development perspective enables the communicative planning process to combine a long-term goal with short-term measures. This is beneficial for most of the stakeholders and creates stable and linear development processes. Flexible adaptation planning, which can cope with uncertainty, goes one step further. Also referred to as swarm planning, flexible adaptation planning tries to anticipate unexpected future changes and proposes a flexible network approach to planning in such a way that adaptations can be implemented in an urban region before the climate change event occurs, outpacing uncertainty. In terms of complexity, swarm planning includes crises, allowing a complex system to move from one attractor to another, unpredictable, attractor, or even to initiate the move. There is a limit to certainty in planning processes; it is not possible to know everything (Cilliers 2005; Byrne 2005). Spatial planning contributes to a better understanding of changes that take place in complex adaptive systems, but, by definition, every complex system will have to deal with uncertainty and new and unforeseeable events.

3.7 Conclusion

This chapter has identified the complexity concepts that are of use in describing and using spatial planning for climate change adaptation. Since climate change is, by definition, a process of transformation, the question under which conditions a complex system is likely to change to a higher level of complexity, acquiring a higher adaptive capacity, is highly relevant in this respect. Concepts from Complexity Theory have been used to develop a framework in which these changes can be identified and explained. Four distinct planning strategies (planning for mitigation, sectored adaptation, integrated adaptation and flexible adaptation) have been positioned within this framework, illustrated by practical planning examples from various European countries. We can conclude that each strategy has its own niche in the framework, that all four together describe the behaviour of a complex system as a whole (Fig. 3.11) and that flexible adaptation planning is the most suitable option to enhance system change. When this reasoning is

Fig. 3.11 Woldendorp: typical landscape of the Eemsdelta region, including small villages built on artificial hills, close to the sea

reversed and the question is posed which planning strategy fits in best with the demands imposed by climate change (e.g. the need for system change), flexible adaptation planning is seen as the most suitable option.

References

Adger WN (2003) Social capital, collective action and adaptation to climate change. Econ Geogr 79(4):387–404

Anderson P (1999) Complexity theory and organisation science. Organ Sci 10(3):216–232

Artigani R (2005) Leadership and uncertainty; complexity and the lessons of history. Futures 37:585–603

Axtell R, Epstein J (1996) Growing artificial societies: social science from the bottom up. Brookings Institution, Washington, DC

Bedsworth LW, Hanak E (2010) Adaptation to climate change. J Am Plan Assoc 76(4):477–495
Brace K, Geoghegan H (2010) Human geographies of climate change: landscape, temporality, and lay knowledges. Prog Hum Geogr 35(3):284–302
Byrne D (2003) Complexity Theory and Planning Theory: a necessary encounter. Plan Theory 2(3):171–178
Byrne D (2005) Complexity, configuration and cases. Theory Cult Soc 22:95–111
Campbell H (2006) Is the issue of climate change too big for spatial planning? Plan Theory Pract 7(2):201–230
Charmaz K (2006) Constructing Grounded Theory: a practical guide through qualitative analysis. Sage Publishers, London
Cilliers P (2005) Complexity deconstruction and relativism. Theory Cult Soc 22:255–267
Cohen J, Stewart I (1994) The collapse of chaos. Discovering simplicity in a complex world. Viking, New York
Commonwealth of Australia (2007) Tackling wicked problems; a public policy perspective. Australian Government/Australian Public Service Commission, Canberra
Crawford TW, Messina JP, Manson SM, O'Sullivan D (2005) Guest editorial. Environ Plan B 32:792–798
De Jonge J (2009) Landscape architecture between politics and science, an Integrative perspective on landscape planning and design in the network society. PhD thesis, Wageningen University/Uitgeverij Blauwdruk/Techne Press, Wageningen/Amsterdam
De Roo G, Porter G (2007) Fuzzy planning, the role of actors in a fuzzy governance environment. Ashgate Publishing, Aldershot/Burlington
Duit A, Galaz V (2008) Governance and complexity – emerging issues for governance theory. Gov Int J Policy Adm Inst 21(3):311–335
Faludi A (2004) The impact of a planning philosophy. Plan Theory 3(3):225–236
Ford JD, Berrang-Ford L, Paterson J (2011) A systematic review of observed climate change adaptation in developed nations. Clim Change 106:327–336
Füssel HM (2007) Adaptation planning for climate change: concepts, assessment approaches, and key lessons. Sustain Sci 2:265–275
Geldof GD (2001) Omgaan met complexiteit bij integraal waterbeheer. PhD thesis, University of Twente, TAUW, Deventer
Geraldi JG (2008) The balance between order and chaos in multi-project firms: a conceptual model. Int J Proj Manag 26:348–356
Glaser B, Strauss A (1967) The discovery of Grounded Theory: strategies for qualitative research. Aldine, Chicago
Gleick J (1987) Chaos: making a new science. Viking/Penguin, New York
Gornitz V, Couch S, Hartig EK (2002) Impacts of sea level rise in the New York City metropolitan area. Glob Planet Change 32:61–80
Grimm NB, Faeth SH, Golubiewski NE, Redman CL, Wu J, Bai X, Briggs JM (2008) Global change and the ecology of cities. Science 319:756–760
Healy P (1997) Collaborative planning – shaping places in fragmented societies. Macmillan Press, London
Healy P (2003) Collaborative planning in perspective. Plan Theory 2(2):101–123
Hillier J (2008) Plan(e) speaking: a multiplanar theory of spatial planning. Plan Theory 7(1):24–50
Hunt A, Watkiss P (2011) Climate change impacts and adaptation in cities: a review of the literature. Clim Change 104:13–49
Innes JE (2004) Consensus building: clarification for the critics. Plan Theory 3(1):5–20
Kauffman S (1993) The origins of order. Oxford University Press, New York
Kelly J, Stark D (2002) Crisis, recovery, innovation: responsive organization after September 11. Environ Plan A 34:1523–1533
Lansing JS (2003) Complex adaptive systems. Annu Rev Anthropol 32:183–204
Lazarus R (2009) Super wicked problems and climate change: restraining the present to liberate the future. Cornell Law Rev 94:1053–1233

Levinthal DA, Warglien M (1999) Landscape design: designing for local action in complex worlds. Organ Sci 10(3):342–357

Lewin R (1992) Complexity: life at the edge of chaos. Macmillan, New York

Liu J, Dietz T, Carpenter SR, Alberti M, Folke C, Moran E, Pell AN, Deadman P, Kratz T, Lubchenco J, Ostrom E, Ouyang Z, Provencher W, Redman CL, Schneider SH, Taylor WW (2007) Complexity of coupled human and natural systems. Science 316:1513–1516

Lorenz EN (1963) Deterministic non-periodic flow. J Atmos Sci 20:130–141

Lynas M (2008) Six degrees. Harper Collins Publishers, London

Mason RB (2007) The external environment's effect on management and strategy, a complexity theory approach. Manag Decis 45(1):10–28

Manson S, O'Sullivan D (2006) Complexity theory in the study of space and place. Environ Plan A 38:677–692

Mitchell Waldrop M (1983) Complexity, the emerging science at the edge of order and chaos. Viking Books, London

Montalvo C (2006) What triggers change and innovation? Technovation 26:312–323

Nath KP, Behera B (2011) A critical review of impact of and adaptation to climate change in developed and developing countries. Environ Dev Sustain 13:141–162

Nowotny H (2005) The increase of complexity and its reduction: emergent interfaces between the natural sciences, humanities and social sciences. Theory Cult Soc 22:15–31

O'Connor MK, Netting FE, Thomas ML (2008) Grounded Theory: managing the challenge for those facing Institutional Review Board oversight. Qual Inq 14(1):28–45

O'Sullivan D (2004) Complexity science and human geography. Trans Inst Br Geogr NS 29:282–295

O'Sullivan D, Manson SM, Messina JP, Crawford TW (2006) Guest editorial. Environ Plan A 38:611–617

Orlove B (2005) Human adaptation to climate change: a review of three historical cases and some general perspectives. Environ Sci Policy 8:589–600

Plowman DA, Baker LT, Beck TE, Kulkarni M, Solansky ST, Travis DV (2007) Radical change accidentally: the emergence and amplification of small change. Acad Manag J 80(3):515–543

Portugali J (2006) Complexity theory as a link between space and place. Environ Plan A 38:647–664

Preston BJ, Westaway RM, Yen EJ (2011) Climate change adaptation in practice: an evaluation of adaptation plans from three developed nations. Mitig Adapt Strateg Glob Change 16:407–438

Prigogine I (1986) Civilisation and democracy: values, systems, structures and affinities. Futures 18:493–507

Prigogine I, Stengers I (1984) Order out of chaos. Bantam Books, New York

Pulselli RM, Tiezzi E (2009) City out of chaos. WIT Press, Southampton

Raschky PA (2008) Institutions and the losses from natural disasters. Nat Hazard Earth Syst Sci 8:627–634

Richards A (2002) Complexity in physical geography. Geography 87(2):99–107

Roggema R (2008) The use of spatial planning to increase the resilience for future turbulence in the spatial system of the Groningen region to deal with climate change. In: Proceedings UKSS-conference, Oxford

Roggema R (2009a) Regional planning for a changing climate in the province of Groningen. In: Van den Dobbelsteen A, van Dorst M, van Timmeren A (eds) Smart building in a changing climate. Techne Press, Amsterdam, pp 47–62

Roggema R (2009b) Adaptation to climate change, does spatial planning help? Swarm Planning does! In: Brebbia CA, Jovanovic N, Tiezzi E (eds) Management of natural resources, sustainable development and ecological hazards. WIT Press, Southampton, pp 161–172

Roggema R (2010) Landscape architecture and climate adaptation: new chances for spatial identity. Landsc Archit China 11:24–31

Roggema R (2011) Swarming landscapes, new pathways for resilient cities. In: Proceedings 4th international urban design conference. Resilience in urban design, Surfers Paradise

Roggema R, van den Dobbelsteen A (2008) Swarm planning: development of a new planning paradigm, which improves the capacity of regional spatial systems to adapt to climate change. In: Proceedings world sustainable building conference (SB08), Melbourne

Roggema R, van den Dobbelsteen A, Biggs C, Timmermans W (2011) Planning for climate change or: how wicked problems shape the new paradigm of Swarm Planning. In: Proceedings World Planning Schools conference, Perth

Ruth M, Coelho D (2007) Understanding and managing the complexity of urban systems under climate change. Clim Policy 7:317–336

Sanchez-Rodriguez R (2009) Learning to adapt to climate change in urban areas. A review of recent contributions. Curr Opin Environ Sustain 1:201–206

Scheffer M, Carpenter S, Foley JA, Folke C, Walker B (2001) Catastrophic shifts in ecosystems. Nature 413:591–596

Strauss AL, Gorbin J (1998) Basics of qualitative research: techniques and procedures for developing Grounded Theory, 2nd edn. Sage, Thousand Oaks

Teisman GR, Klijn EH (2008) Complexity theory and public management. Public Manag Rev 10(3):287–297

Tiezzi E (2003) The essence of time. WIT Press, Southampton/Boston

Timmermans W (2004) Crisis and innovation in sustainable urban planning. In: Advances in architecture, vol 18. WIT-Press, Southampton, pp 53–63

Timmermans W (2009) The complex planning of innovation. In: Transactions on ecology and the environment, vol 122. WIT-Press, Southampton, pp 581–590

Timmermans W, van Dijk T, van der Jagt P, Onega Lopez F, Crecente R (2011) The unexpected course of institutional innovation processes: inquiry into innovation processes into land development practices across Europe. Int J Des Nat Ecodyn 6(4):297–317

Underdal A (2010) Complexity and challenges of long term environmental governance. Glob Environ Change 20:386–393

Urry J (2006) Complexity. Theory Cult Soc 23:111–115

Vicenzi R, Adkins G (2000) A tool for assessing organizational vitality in an era of complexity. Tech Forecast Soc Change 64:101–113

VROM-raad (2007) De hype voorbij, klimaatverandering als structureel ruimtelijk vraagstuk. Advies 060. VROM-raad, Den Haag

Wilson E (2007) Adapting to climate change at the local level: the spatial planning response. Local Environ 11(6):609–625

Wheeler SM (2008) State and municipal climate change plans: the first generation. J Am Plan Assoc 74(4):481–496

Woestenburg M, Kuypers V, Timmermans W (2006) Over de brug, hoe zanderij Craailo een natuurbrug kreeg. GNR, Hilversum. Blauwdruk, Wageningen

Websites

www.factsproject.eu
http://www.factsproject.eu/pilotprojects/dewijers/Pages/default.aspx
http://www.factsproject.eu/pilotprojects/varna/Pages/default.aspx
http://www.factsproject.eu/pilotprojects/baixovougalagunar/Pages/default.aspx

Chapter 4
Transition and Transformation

Rob Roggema, Tim Vermeend, and Wim Timmermans

Contents

4.1	Introduction	68
	4.1.1 Resilience	68
	4.1.2 Change in Current Spatial Planning	69
4.2	Transition	70
	4.2.1 Three Horizons of Change	70
	4.2.2 Transition Phases	70
	4.2.3 A Slow Pace or Advanced Transition	74
4.3	Transformation	75
4.4	B-Minus	81
4.5	Early Signals	84
	4.5.1 Early Warning Signals	85
	4.5.2 Creation of Starting Points for Change	86
4.6	Conclusion	87
References		88

Abstract In this chapter it is argued that fundamental change in society is required, because environmental problems are serious and ask for a factor 10 or more shift in society, the resilience approach (as outlined in Chap. 2) implies change to higher

---

R. Roggema (✉)
The Swinburne Institute for Social Research, Swinburne University of Technology,
PO Box 218, Hawthorn, VIC 3122, Australia
e-mail: rob@cittaideale.eu

T. Vermeend
UC Architects, Guyotplein 5, 9712 NX, Groningen, The Netherlands
e-mail: timvermeend@ucarchitects.com

W. Timmermans
Groene Leefomgeving van Steden, Wageningen UR – Van Hall Larenstein – Tuin en Landschapsinrichting, PO Box 9001, 6880 GB, Velp, The Netherlands
e-mail: wim.timmermans@wur.nl

R. Roggema (ed.), *Swarming Landscapes: The Art of Designing For Climate Adaptation*,
Advances in Global Change Research 48, DOI 10.1007/978-94-007-4378-6_4,
© Springer Science+Business Media Dordrecht 2012

resilience systems and current spatial planning is unable to facilitate fundamental change. Transition of an existing system into a better version of the same system does not comply with the demands of fundamental changes. Instead of choosing for the pathway of change, a change of pathway is required. This transformation of the existing stable regime (system A) into a fundamental other regime (system B) is able to meet the urgency to change. However, Transformation of a system is only possible when the new system is fundamental separated from the original and is capable to develop its own growth curve. The proposed pathway courses via B-minus. A predecessing state of system B consisting of rudimentary spatial elements, which can be observed as critical early warning signals and can be created at specific intersections in the network. These signals require a spatial translation to become useful in spatial planning. Network analysis is needed to determine the locations where to create starting points for a system change.

Keywords Transformation • Transition • Spatial planning • Networks • Early warning signals

4.1 Introduction

"Contemporary environmental problems, such as climate change, loss of biodiversity and resource depletion, present formidable societal challenges. Addressing these problems requires factor 10 or more environmental performance, which can only be realised by deep-structural changes. These systemic changes are often called socio-technological transitions" (Geels 2011). These 'system errors', as Rotmans calls them are "flaws in our societal system, which cannot be corrected through market mechanisms: weak networks, fixation on technology, ingrained behaviour, institutional constraints and path dependencies" (Rotmans 2005). In dealing with uncertainty the resilience approach is, as pointed out in Chap. 2, found valuable, e.g. a certain area or system needs to increase its resilience in order to deal with uncertain circumstances. Both, environmental problems and resilience imply fundamental change. Many elements, undergoing these changes are embedded in spatial planning, plans or processes. However, the way these changes can be reached depends largely on the approach. Is the process seen as a pathway of change, a gradual smooth change, a transition process is useful, but if a change of pathway is required a transformation is more obvious. Both approaches are addressed in this chapter.

4.1.1 Resilience

The concept of resilience (Walker et al. 2004) has mainly been used to study socio-ecological systems (amongst other: Olsson et al. 2006; Wilkinson et al. 2009; Cork 2010). Notwithstanding the major contribution these works deliver in understanding the resilience of the Earth system and other socio-ecological systems, the (potential increase of) resilience of spatial systems such as cities and landscapes is less extensively studied.

The Earth system (Lenton and Van Oijen 2002; Lovelock 1988) and spatial systems, such as landscapes and cities are defined as complex adaptive systems (Allen 1996; Portugali 2000; Batty 2005), which in principle make them suitable for resilience thinking. Moreover, complexity is increasingly seen as a fundamental theory for spatial planning (Innes and Booher 2010; De Roo and Porter 2007).

The adaptability, e.g. the collective capacity of actors in the system to manage resilience (Walker et al. 2004), of spatial systems is determined by the collective capacity of spatial elements to manage resilience. Here, spatial elements are defined following Dalton and Bofna (2003): "Elements of zero, one, and two dimensions that observers acquire and utilize as anchors for location (…). Not only can the observer position himself in space in terms of basic topological relationships ('to the front of', or 'to the right of') but also 'at', 'on', or 'inside' them". Hence, spatial elements have the collective capacity to manage resilience, which allows spatial systems to increase its resilience.

4.1.2 Change in Current Spatial Planning

Spatial planning practice has major difficulties to facilitate fundamental change, as it is not used to major shifts and changes. This can be illustrated using the three most recent regional plans (Provincie Groningen 2000, 2006, 2009) for the Groningen province area in the Netherlands. The changes in aims, policies, chapters, and maps are marginal. Once policies are defined in the first plan they are repeated to a large extent in the second and third plan. When the functional maps of the first and third plan are compared a modest 2% of the entire area is allowed to undergo any functional change over a period of 13 years. This example illuminates the 'incre*mentality*' that is manifest in many spatial plans, at least in the Netherlands. These incremental changes in the consecutive regional plans can be visualised as a straight, slowly rising line on which identical waves of planning processes follow each other (Fig. 4.1).

The small changes that are the result of these consecutive spatial plans do not meet the needs for fundamental changes. A preliminary design, in which the required changes to adapt to climate change are integrated shows that approximately 30% of the land area needs to potentially undergo a functional change (Roggema 2007), far more than the 2% that is allowed.

Given the required changes, as a result of the type of problems society faces and as a result of striving for higher resiliency, and the inability of current spatial planning practice to incorporate change, the search for a fundamental new planning approach is necessary. This new planning framework (Roggema et al. 2012), is capable of identifying the required changes and will be elaborated on in Chaps. 6 and 7 of this book. The question, once we know what we want to achieve, how to reach this changed future is discussed in this chapter.

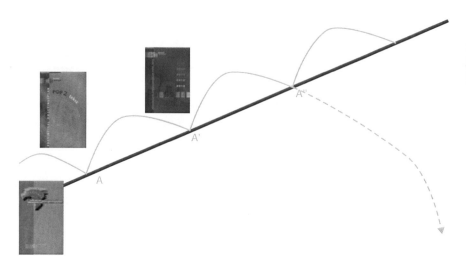

Fig. 4.1 Subsequent plans of the same 'family' as consecutive waves

4.2 Transition

4.2.1 Three Horizons of Change

Once it becomes clear that a planning approach, which operates in terms of end images for the future is not suited to resolve the longer-term dynamics of climate change and energy supply, alternative pathways are to be explored. These alternative pathways do not take a certain end-result as the main focus, but the process or the transition towards an uncertain future. However, before this pathway can be determined the question is which uncertain future we want. As demonstrated by Newton (2008) in his 3-horizons model, the more sustainable futures take a longer period to implement (Fig. 4.2) and as such determine the pace and path of the transition.

4.2.2 Transition Phases

In recent literature the change from a certain state or regime towards another (more sustainable) is described as a transition. A transition is defined as "a gradual, continuous process of societal change, changing the character of society (or a complex part) structurally" (Rotmans et al. 2000). This transition is generally represented by a fluent curve (Fig. 4.3) and divided into four phases: pre-development, take-off, acceleration and stabilisation.

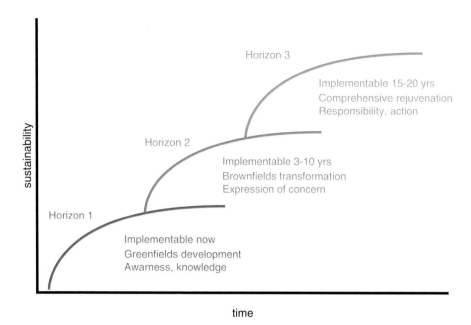

Fig. 4.2 3-Horizon thinking (After Newton 2008)

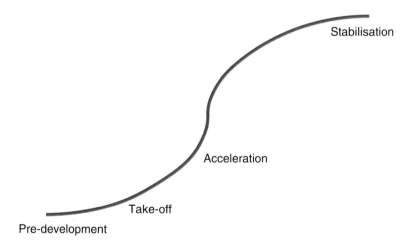

Fig. 4.3 Basic phases of a transition (After Rotmans et al. 2000)

De Roo (2008) elaborates on this and attributes dynamics to each of the distinguished phases (Fig. 4.4): between the two stable phases a dynamic phase enables the system to shift from an old (weak) context towards a new (stronger) one.

Various studies on change management argue that this change can only take place if a crisis has been experienced (Hurst 1997; Peters and Wetzels 1997; Homan 2005; Zuijderhoudt 2007). Corresponding schemes all describe this transition as a

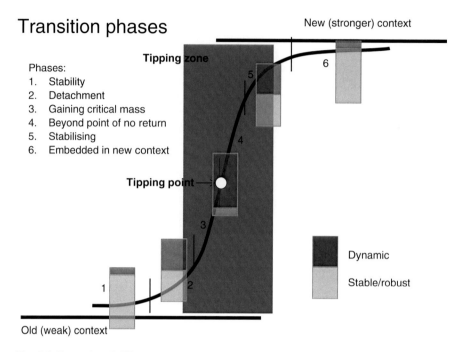

Fig. 4.4 Dynamics of different transition phases (Source: De Roo 2008)

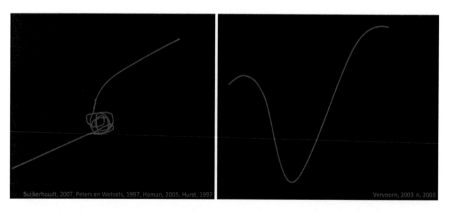

Fig. 4.5 A crisis before reaching a new level (Source: Zuijderhoudt 2007; Hurst 1997; Peters and Wetzels 1997; Homan 2005 (*left*), Vervoorn 2003 (*right*))

fluent line up to a certain point where chaotic circumstances appear. Out of this chaos a new fluent line emerges (Fig. 4.5). Moreover, during times of change, there is no such thing as a quick fix, as Vervoorn (2003) demonstrates, using Dante's *Divina Commedia* as a metaphor (Fig. 4.6 and Box text): society needs to create a very clear, imaginable and attractive image of the future vision before it can leave the old behind and learn, while experimenting the new.

Box Text

The narrative Hans Vervoorn uses to explain transition, tells the story of Dante Algieri seeking for his beloved Beatrice: "Dantes Divina Commedia describes the process that change only can take place if a crisis has been experienced before. In the middle of my life, Dante writes in 1315 in the first paragraph of chapter one in La Divina Commedia, I lost track and ended up in a fearful overgrown and dark wood. In our times we would say: I am in deep trouble. Somewhere, far away, Dante sees a sunny hill, where he would like to jump to. Nowadays we would say: you're in denial, because there is seldom a quick fix. It is impossible for Dante to jump to the hill, because three wild animals are making trouble. These animals represent the three basic human fears for change:

(a) The lion stands for pride. The basic fear is ego hurting: the fear that, in case of change, you are accused you did things wrong before.
(b) The panther stands for flexibility. The basic fear is that you are thrown out of your comfort zone in case of change. The fear that you need to do things you never did before and are not yet capable of.
(c) The wolf stands for greed. The basic fear is that you are threatened in your status, position or welfare. In great fear he calls God for help, who sends, as a Deus Ex Magina, Vergil, a poet in the early Middle Ages, for whom Dante has great respect. Vergil says to Dante: I have some good and bad news for you:

The good news is: I will get you out of this dark forest and will bring you to Beatrice (a very beautiful girl from Firenze, with whom Dante fell in love with in his youth and wrote many love-poems about) and with her you will reach Heaven, the dream of every Christian.

The bad news is that our pathway will lead us through hell (a metaphor for de-learning, get rid of the old) and purgatory (learning the new while experimenting).

In de rest of his book Dante describes exactly this pathway. The moral of this story is:

In times of change you need a coach, an advisor (Vergil) who helps Dante (society) during the process. The first thing society (Dante) needs to create is a very clear, imaginable, attractive image of the vision, the ultimate dream and final image (Beatrice), before society can overcome its basic fears, leave the old (Hell) behind and learn (Purgatory), while experimenting the new." (Vervoorn 2003)

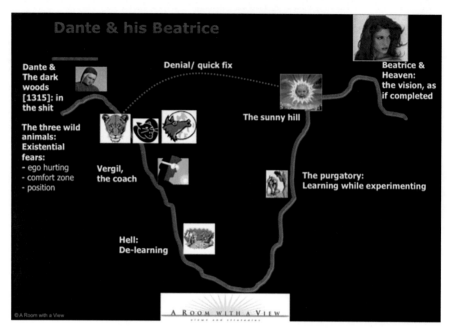

Fig. 4.6 Dantes Divina Commedia as metaphor for change management (Source: Vervoorn 2003)

The fluency of the line in Figs. 4.5 and 4.6 represents the fact that the system itself is not fundamentally changed. After the transition the same system has reached a new stable state of a higher quality.

4.2.3 *A Slow Pace or Advanced Transition*

In general a transition starts when some sense of urgency is felt. For example, in the peak-oil case the transition is starting to 'take-off' very slowly, due to the awareness of urgency amongst a large group of people. While oil reserves are shrinking rapidly at the same time, the question is whether the transition to a system functioning without oil can be completed before we run out of fossil resources: a disaster of leaving a large number of people without energy. The transition starts at the moment the *zone of urgency* (Fig. 4.7) is entered, but is only completed after the disaster has happened. Problem solved, but the disaster could not be prevented.

Two ways of alternative, more anticipative, transitions are distinguished: (I) an *advanced* and (II) a *slow-pace* transition (Fig. 4.7).

In order to prevent the disaster from happening an alternative transition pathway needs to be developed. The first alternative is to keep the pace of the transition the

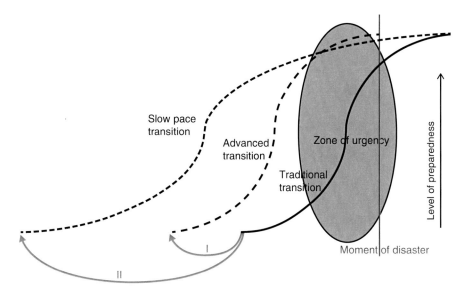

Fig. 4.7 Advanced (I) and Slow pace (II) transition

same, but begin the process earlier: an *Advanced* transition (I). The early start allows the transition to be completed sooner. Problem here is that required changes are needed in the same pace, but without a sense of urgency, which makes the reason for change unclear. Another alternative is found to start the transition-process earlier, but also 'down-pace' the speed of the transition: a *Slow-pace* transition (II). This makes it possible to implement small steps of change that are acceptable without the sense of urgency. Both pathways have their transition completed before entering the zone of urgency and ahead of the moment a disaster occurs.

4.3 Transformation

Transformation trajectories are the subject of a growing body of literature (Gunderson and Holling 2002; Geels and Kemp 2006; Chapin et al. 2010). Burgess defines transformation as "the ability to change to a new identity if the old one is not appropriate" (Burgess 2010). Folke and colleagues (2010) describe a transformation as "the capacity to transform the stability landscape itself in order to become a different kind of system, to create a fundamentally new system when ecological, economic, or social structures make the existing system untenable". A "fundamental change in a social–ecological system results in different controls over system properties, new ways of making a living and often changes in scales of crucial feedbacks" (Chapin et al. 2009). Transformations can be purposefully navigated or happen unintended.

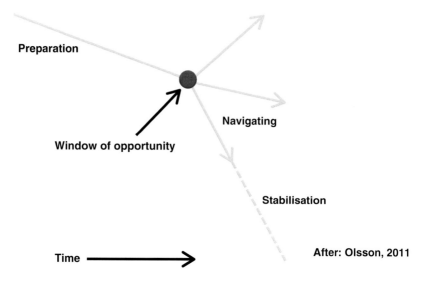

Fig. 4.8 Three phases of transformation (After: Olsson 2011)

In the works of Olsson and colleagues (2006) and Chapin and colleagues (2009), transformation is divided in three phases (Fig. 4.8): preparing, navigating and stabilising. The moment between preparing for and navigating a transformation is defined as the window of opportunity (Olsson et al. 2006). The property of a window of opportunity is "the presence of many options yet a short time-frame to start a transformation" (Olsson et al. 2006), the "occurrence of a crisis, which makes it possible to plan for a transformation" (Chapin et al. 2009) and a "set of activities pursed" (Olsson et al. 2006) "in a sequence of events leading to the start of navigating the transformation" (Olsson 2011).

Despite the fact that this process is called a transformation, it may be questioned whether this described change is (limited to) only a change of direction within one system, and making it better prepared for changed circumstances. Hence, it does not describe a transformation of the system into another system.

Blauwhof and Verbaan (2009), based on Perez (2002), argue that subsequent (and disconnected) 'waves' appear and that the 'next' wave has already started while the former is still ongoing (Fig. 4.9). The disconnected waves 'overlap' within a certain zone (A and B in Fig. 4.9), which operates as the window of opportunity, where navigation a transformation starts.

Ainsworth-Land defines "Growth as the single process (in nature) that forms the keystone of transformation theory and that unites the behaviour of all things". He distinguished three phases of growth: forming (Phase 1), norming (Phase 2) and integrating (Phase 3). These three phases together shape the transformation. In between successive growth cycles (the transformations) a stable, growthless, period occurs. This transition period is therefore represented through a flat line (Fig. 4.10). At a later stage he adjusted his theory and defined the transformation periods as overlapping growth cycles (Fig. 4.11). A new phase one (forming) starts already while phase three (integrating) of the previous cycle is still active (Ainsworth-Land 1986).

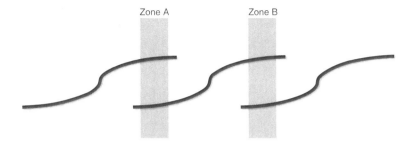

Fig. 4.9 Disconnected 'waves' (Source: Blauwhof and Verbaan)

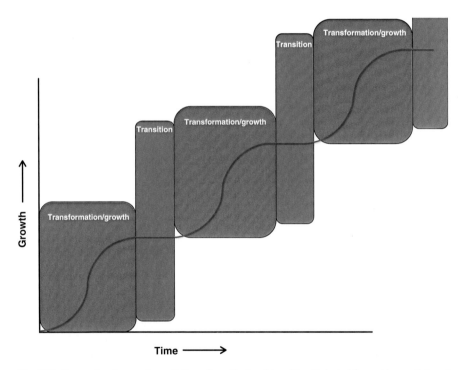

Fig. 4.10 Successive phases of growth (transformation) and transition (Adapted from: Ainsworth-Land 1986)

This new phase one interfering the existing growth cycle or regime originates through the development of niche innovations, one of the levels that are part of the multi-level perspective (Geels 2002, 2005, 2011). The multi-level perspective theorises non-linear processes resulting from an interplay of developments on three analytical levels: niches, (the locus of radical innovations), socio-technical regimes (the locus of established practices and associated rules that stabilise existing systems) and the

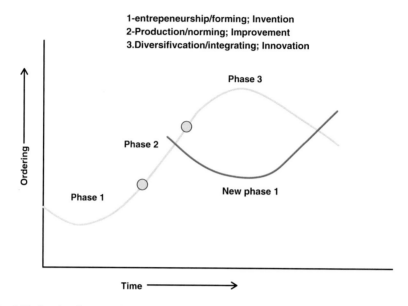

Fig. 4.11 Overlapping growth cycles (Ainsworth-Land 1986)

exogenous socio-technical landscape, representing the nearly unchangeable values and biophysical features of the system. The level of the socio-technical regime is the level that is stable and the level where change is effective on, because here the regime can shift from one to another. Change starts in niches or where novel configurations appear (Geels 2002). The effectiveness of the change e.g. weather a regime shift will occur depends on the alignment of developments. Successful processes within the niche are reinforced by changes at the regime and/or the landscape level (Kemp et al. 2001: 277). Hence, this reinforcement determines whether a novelty fails, modifies the regime or transforms the landscape (Fig. 4.12).

The process of change consists of several elements (Fig. 4.13). The existing regime is dynamically stable (point 2), which means that it is potentially open for change. However, it will only open up if the pressure from the landscape level creates a window of opportunity (point 1). Both levels then influence externally the niches (point 3, 4), which supports the development of novelties (point 5). Once these novelties are developed and are aligned towards a dominant design (point 6), they are capable of breaking through the existing regime (point 7) and enforce adjustments to the old regime, which then will transform into a new regime. Eventually, when the regime shifts are profound, they may influence the landscape level, changing the set of values and/or biophysical properties (point 8).

Elaborating on the former theories, the transformation of a system originates somewhere outside the existing regime or system while the system is still operating in its dynamically stable regime. The start of phase one (forming mode) of the growth curve of system B takes place where niche innovations are located, while system A (the stable regime) is still functioning in its integrating mode. The forming of system B only takes place through novelty development, disconnected from the

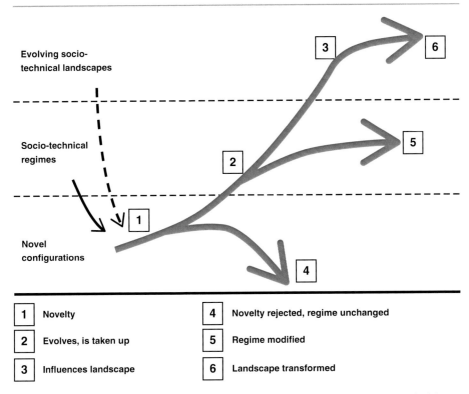

Fig. 4.12 Uptake of novelties (Adapted from Rip and Kemp 1996; Kemp et al. 2001; cited in: Geels 2002)

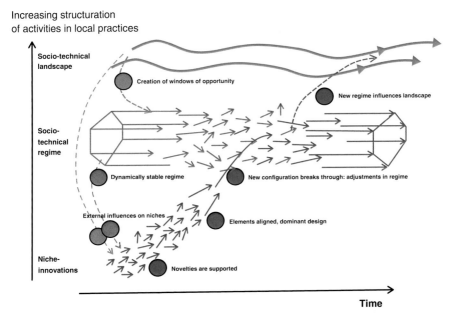

Fig. 4.13 Interaction between the levels of the multi-level perspective (Geels 2002, 2005, 2011)

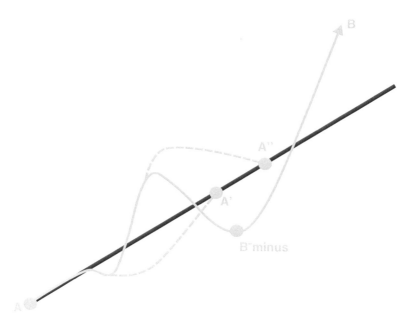

Fig. 4.14 The fluent line of transition changes A in A-**apostrophe**, while the shift to B requires a discontinuous process through B-**minus**

current stable regime. Whenever these novelty developments return to the stable regime and become part of the existing again, an adjusted system A has been created (e.g. *A-apostrophe* or *A-double apostrophe*, Fig. 4.14). System B, represented through the novelty developments, needs to follow its own growth curve of forming, whilst crossing system A, and overtaking it in its norming and integrating phases (Fig. 4.14). Here, we name the forming phase of B, *B-minus*.

In order to ignite the forming of system B the transformability, "the capacity to create a fundamental new system, when ecological, economical or social conditions make the existing system untenable" (Walker et al. 2004), of the system must be 'triggered' by a new and attractive vision on what this alternative system (B) has to offer. Only then the system will change pathways, transforming, and derails from the pathway of change (transition). Transition brings an existing system out of balance and into a new stable state of the same system, while a transformation transforms the system into a fundamentally new one: the transformation (growth) of B replaces system A. A resilience approach allows a new identity to emerge through interactions within and across scales, introducing new defining state variables and losing others (Folke et al. 2010). Transformations are announced through elements that represent the forming of system B, although they are invisible yet.

1. If fundamental shifts in socio-economic, technological and spatial systems are required a transformation is suitable;
2. Transformations describe a fundamental shift;
3. Both transitions as transformations can be used to define and achieve an attractive future (B) in the face of climate change. However, the change

implied through a transformation might be larger than through a transition might be reached.

A transformation process offers better prerequisites to deal with and achieve the fundamental change.

4.4 B-Minus

The concepts of transformation are elaborated and new concepts were developed during the so-called pizza debates; e.g. in small group sessions of (2–3) people the concept of transformation has been in depth discussed. Over a period of 6 months regularly meetings took place and an iterative process of consecutive 'brainstorm-capture-writing-brainstorm-elaboration-capture-writing' was organised. This has led to the description of the process of transformation, starting in B-minus.

Current spatial plans include changes that are too small to call it fundamental change. This is caused by the fact that during a period of stability and satisfaction existing paradigms continue to be adapted in political cycles. After a while, political cycles come to an end and allow for a shift to a fundamental new paradigm. The repetitive spatial plans are improved during this stable period, but they stay within the same type. A shift from A (the original) towards *A-apostrophe* and *A-double apostrophe* takes place, but B, a fundamental new type of spatial plan will never be achieved (Fig. 4.15).

Every transformation needs to start with framing the desired future (system B) in an attractive way that responds to a certain urgency. Elements such as icons, identity (Castells 1996), branding (Franzen and Bouwman 1999; Roberts 2006), branding identity (Ghodeswar 2008), and a stickiness factor (Gladwell 2000), all play an important role in making the future vision attractive. Moreover, dynamic planning has to be applied when an attractive climate adaptive future needs to be designed (Berger and Chambwera 2010). Only then, high expenses and existing standards can be overcome.

We distinguish two situations, in which a change of pathway towards is likely to occur.

1. A disaster can disturb the stable regime, represented as a regular, straight pathway of steps towards the future (Fig. 4.16). In case of a disaster regular policy will temporarily no longer be relevant, as immediate action is required. An instant a shift from one pathway to another is likely. Hardly visible to regular policy-making, pathway B was already in operation but suddenly becomes interesting, as it enables pathways to recover and provides solutions for the longer term (Fig. 4.17).
2. The current system slowly fades away, for instance because it does not meet current demands anymore. At a certain point another system takes over (Fig. 4.18), because the new system (B) contains the features the current timeframe demands.

The change from the current functioning system (A) towards a new system (B), induced by a disaster or a slow fade away (Fig. 4.19), takes place through the

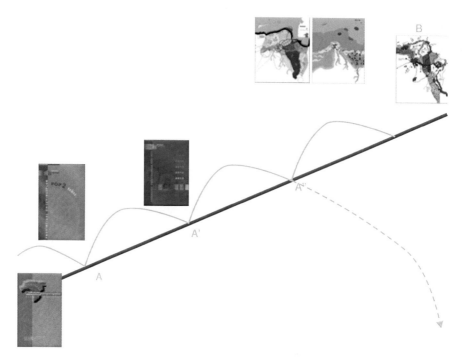

Fig. 4.15 The desired future system (B) is defined, but 'missed' by consecutive spatial plans

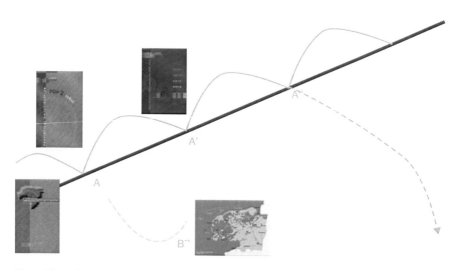

Fig. 4.16 A disaster enforces the move of system A off-track

4 Transition and Transformation

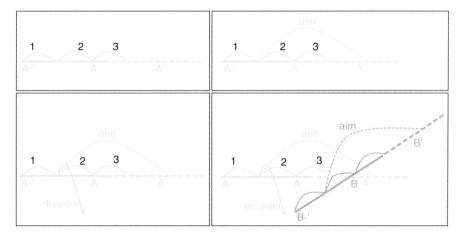

Fig. 4.17 After a disaster pathway B takes over

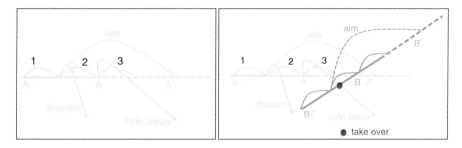

Fig. 4.18 System A fades away and B takes over

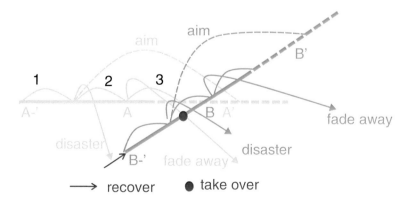

Fig. 4.19 Recovery and fade away in one image

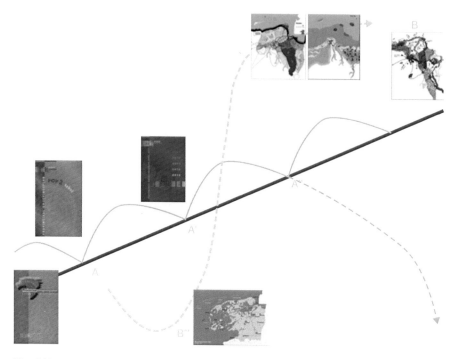

Fig. 4.20 Towards B via B-**minus**

pre-phase of B: '*B-minus*' (Fig. 4.20). This rudimentary stage (the *forming phase*, or the location of *niche innovations*) of the new system B contains elements of the new system, but is far from complete. In order to anticipate or stimulate a transformation it is important to recognise or create these predecessors of B. This will be discussed in the following chapter.

4.5 Early Signals

Once the attractive future system is defined, it is possible to search for the elements of *B-minus*, as these can be seen as a *backtracked* version of that future. In contrast with forecasting (predicting the future starting from present) or backcasting (define the desired future, and derive from that the steps to be taken to realise that future), backtracking goes back in history to find a sustainable equilibrium, which functions as an inspiration for defining a desired future system and from derive from that the steps to realise it (Fig. 4.21) (Schoot Uiterkamp et al. 2005).

The window of opportunity as defined by Olsson et al. (2006) is the key moment to start a transformation and also the moment when elements of the *B-minus* state become visible.

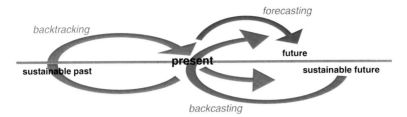

Fig. 4.21 The difference between forecasting, backcasting and backtracking (Van den Dobbelsteen et al. 2006)

The elements of *B-minus* can be determined in two ways:

1. Observation or active search for signals announcing a transformation, so-called early warning signals (Scheffer et al. 2009);
2. The active creation of harbingers of a transformation.

4.5.1 Early Warning Signals

In the work of Scheffer et al. (2009) early warning signals are defined for systems approaching a major change. Despite the fact that it is very difficult to develop accurate models to predict thresholds in most complex systems, Scheffer and colleagues discuss the generic character of early warning signals from a range of complex systems. They conclude: "if we have reasons to suspect the possibility of a critical transition, early-warning signals may be a significant step forward when it comes to judging whether the probability of such an event is increasing". They distinguish the following signals:

1. Critical slowing down: The intrinsic rates of change in the system decrease, leading to a system state that more and more resembles its past state. Two symptoms are distinguished: increase of autocorrelation and increase of variance.
2. Skewness: An unstable equilibrium, which marks the border of the basin of attraction, approaches the attractor from one side. In the vicinity of this unstable point the rates of change are lower. As a result, the system will tend to stay in the vicinity of the unstable point relatively longer.
3. Flickering: The system moves back and forth between the basins of attraction of two alternative attractors.
4. Types of spatial patterns: (1) scale-invariant distributions of patch sizes and increased spatial coherence, or (2) the appearance of regular patterns in systems governed by local disturbances.

These signals not necessarily contain a spatial dimension or make them easy to use or understand in a spatial planning context. However, in Table 4.1 a first attempt to 'translate' the early warning signals into possible spatial dimensions is presented.

Table 4.1 Translation of early warning signals into spatially relevant dimensions

Announcement of system change (early warning signals, derived from Scheffer et al. 2009)	Possible translation into spatial dimensions
Critical slowing down (increase of autocorrelation, increase of variance)	Maintaining old historic structures, re-emphasize existing patterns of functions
	Repetitive policies (the longer policies remain unchanged or are repeated over and over again, the closer we are to a system change)
Skewness	Dominance of one centre over another, core-periphery
Flickering	Temporarily repetitive occupation for living, temporarily repetitive flooding
1. Scale-invariant distributions of patch sizes/increased spatial coherence 2. Increase of regular patterns	Urban sprawl, repetitive urban patterns/building blocks

4.5.2 Creation of Starting Points for Change

Besides trying to identify early warning signals, another option is to actively create the starting-points for systems change. Points in networks where developments are likely to start can function as the elements of *B-minus*, places where the niche developments take place and capable of eventually leading us to the new desired system B. Network theory emphasises that some nodes in networks are more suited for the ignition of change than others. The following key characteristics of networks are derived from Newman et al. (2006):

1. *Enough Edges*: Once enough edges are added, properties of the network suddenly increase in quality (Erdós and Rényi 1960);
2. *The Core*: Directed networks consist of a core (a giant, strongly connected component), links-in and links-out, as well as other islands and tendrils, represented visually by Broder et al. (2000) as a bow-tie;
3. *High Level of Clustering*: The small world effect (Watts and Strogatz 1998) describes the characteristics of networks: if the number of nodes in the network increases, while connected by a short path, the total length of paths will increase logarithmically and a high level of clustering will occur. (Castells 1996);
4. *Fitness of Nodes* (Castells 1996): The increase of connectivity of nodes in a network depends on the fitness to compete for links (Bianconi and Barabási 2001). This fitter-gets-richer phenomenon helps to understand the evolution of competitive systems in nature and society;
5. *Connections*: Robust networks, at least complex biological ones, are formed by numerously connected nodes, which are highly clustered and know a minimum distance between any random pair (Solé et al. 2002).

Preliminary research, applying these principles to a concrete spatial situation (Hao and Wang 2010) made the theory useful for spatial planning. This research, in order to determine the points in the network with the greatest potential to start system change, analysed the networks in two ways: (1) the density of individual networks, such as the water- energy- or transport network, and (2) the number of different network types colliding at one physical location. The role of networks in identifying starting points for transformation is elaborated in Chap. 5.

4.6 Conclusion

As discussed in this chapter, current environmental problems require major system changes. In order to facilitate this, two approaches to describe and/or enhance change are investigated: transition and transformation. The main difference between the two is that transition aims to change the system to a better version and transformation emphasises a fundamental change into a new system. When major change is required transformation offers the most suitable way to not only describe the change, but also to stimulate it. Ainsworth-Land describes transformation in terms of growth and acknowledges that the growth of the 'next system' already starts while the current system still flourishes. Geels underpins this through locating the development of novelties separate from the existing stable regime. In this chapter these theories are elaborated, aiming to find the (transformation) pathway to reach this 'next system' (also referred to as system B).

As stated before, the transition of system A leads to a better version of the same system A: *A-apostrophe*. A transformation pathway, leading to fundamentally different (more resilient) system (B) needs to identify the elements that belong to this 'new system'. This transformation therefore starts in *B-minus*, the first (forming) phase of the new system, where the 'preliminary' parts of system B are found.

To determine *B-minus*, two ways are distinguished: through 'discovery' or as 'creation'.

Scheffer and colleagues theorise that early warning systems can be discovered, announcing the approach of a threshold and system change. These early warning signals are found in several types of systems, with exception of spatial systems. The first attempt to define these signals in spatial dimensions is presented in this chapter.

The other way to find *B-minus* elements is to identify the locations where those elements that get system change started are likely to be developed. Learning from network theory, the most dense nodes and the most connected networks are the most likely places.

In comparison, the pathways leading to *A-apostrophe* and B respectively (Table 4.2) have fundamental different properties. The A-apostrophe pathway is useful to enforce change if tame problems in relatively steady environments are to be dealt with. In this case linear thinking and a transition pathway can be used. However, wicked problem in a complex environment benefit from transformational change and non-linear thinking.

Table 4.2 Characteristics of pathways leading to respectively A-apostrophe and B-minus

A-apostrophe	B-minus
Tame problems	Wicked problems
Moderate environments	Turbulent environments
Transition	Transformation
Linear thinking	Non-linear (dynamic) thinking

References

Ainsworth-Land GT (1986) Grow or die. The unifying principle of transformation, Reissuedth edn. Wiley, New York/Chichester/Brisbane/Toronto/Singapore

Allen PM (1996) Cities and regions as self-organising systems, models of complexity. Taylor & Francis, London/New York

Batty M (2005) Cities and complexity, understanding cities with cellular automata, agent-based models, and fractals. The MIT Press, Cambridge, MA/London

Berger R, Chambwera M (2010) Beyond cost-benefit: developing a complete toolkit for adaptation decisions. IIED, London

Bianconi G, Barabási A-L (2001) Competition and multiscaling in evolving networks. Europhys Lett 54:436–442. In: Newman M, Barabási A-L, Watts DJ (eds) (2006) The structure and dynamics of networks. Princeton University Press, Princeton/Woodstock

Blauwhof G, Verbaan W (2009) Wolk 777, over crisis, krimp en duurzaamheid. Uitgeverij Blauwdruk, Wageningen

Broder A, Kumar R, Maghoul F, Raghavan P, Rajagopalan S, Stata R, Tomkins A, Wiener J (2000) Graph structure in the Web. Comput Netw 33:309–320. In: Newman M, Barabási A-L, Watts DJ (eds) (2006) The structure and dynamics of networks. Princeton University Press, Princeton/Woodstock

Burgess T (2010) Reforming Australia's early childhood development system – the role of resilience theory. In: Cork S (ed) Resilience and transformation, preparing Australia for uncertain futures. CSIRO Publishing and ©Australia 21, Collingwood

Castells M (1996) The rise of the network society. Blackwell, Oxford

Chapin FS III, Carpenter SR, Kofinas GP, Folke C, Abel N, Clark WC, Olsson P, Stafford Smith DM, Walker B, Young OR, Berkes F, Biggs R, Grove JM, Naylor RL, Pinkerton E, Steffen W, Swanson FJ (2009) Ecosystem stewardship: sustainability strategies for a rapidly changing planet. Trends Ecol Evol 25(4):241–249

Chapin FS III, McGuire AD, Ruess RW, Hollingsworth TN, Mack MC, Johnstone JF, Kasischke ES, Euskirchen ES, Jones JB, Jorgenson MT, Kielland K, Kofinas GP, Turetsky MR, Yarie J, Lloyd AH, Taylor DL (2010) Resilience of Alaska's boreal forest to climatic change. Can J For Res 40. doi:10.1139/X10-074

Cork S (ed) (2010) Resilience and transformation, preparing Australia for uncertain futures. CSIRO Publishing and Australia 21, Collingwood

Dalton RC, Bofna S (2003) The syntactical image of the city: a reciprocal definition of spatial elements and spatial syntaxes. In: Proceedings 4th international space syntax symposium, London

De Roo G (2008) A theory of transition and its relevance to planning theory and practice. A non-linear understanding of spatial development VIIth meeting of Aesop's Thematic Group on Complexity and Planning, Milano

De Roo G, Porter G (eds) (2007) Fuzzy planning, the role of actors in a fuzzy governance environment. Ashgate, Aldershot/Burlington

Erdós P, Rényi A (1960) On the evolution of random graphs. Publ Math Inst Hung Acad Sci 5:17–61. In: Newman M, Barabási A-L, Watts DJ (eds) (2006) The structure and dynamics of networks. Princeton University Press, Princeton/Woodstock

Folke C, Carpenter SR, Walker B, Scheffer M, Chapin T, Rockström J (2010) Resilience thinking: integrating resilience, adaptability and transformability. Ecol Soc 15(4):20. [online]: http://www.ecologyandsociety.org/vol15/iss4/art20/

Franzen G, Bouwman M (1999) De mentale wereld van merken. Alphen aan den Rijn, Samsom

Geels FW (2002) Technological transitions as evolutionary reconfiguration processes: a multilevel perspective and a case study. Res Policy 31:1257–1274

Geels FW (2005) Processes and patterns in transitions and system innovations: refining the co-evolutionary multi-level perspective. Technol Forecast Soc Change 72:681–696

Geels FW (2011) The multi-level perspective on sustainability transitions: responses to seven criticisms. Environ Innov Soc Trans 1:24–40

Geels FW, Kemp R (2006) Transitions, transformations and reproduction: dynamics of sociotechnical systems. In: McKelvey M, Holmén M (eds) Flexibility and stability in economic transformation. Oxford University Press, New York, pp 227–256

Ghodeswar BM (2008) Building brand identity in competitive markets: a conceptual model. J Prod Brand Manag 17(1):4–12

Gladwell M (2000) The tipping point: how little things can make a big difference. Little, Brown and Company/Time Warner Book Group/New York Back Bay Books, New York

Gunderson L, Holling CS (eds) (2002) Panarchy: understanding transformations in human and natural systems. Island Press, Washington, DC

Hao H, Wang X (2010) Spatial planning for climate change adaptation – test and improve spatial planning methodology for the CCA. Thesis, Van Hall Larenstein, Velp

Homan T (2005) Organisatiedynamica. Sdu uitgevers, Den Haag

Hurst DK (1997) Crisis en vernieuwing, De uitdaging van organisativerandering. Scriptum, Schiedam

Innes JE, Booher DE (2010) Planning with complexity, an introduction to collaborative rationality for public policy. Routledge, London/New York

Kemp R, Rip A, Schot JW (2001) Constructing transition paths through the management of niches. In: Garud R, Karnoe P (eds) Path dependence and creation. Lawrence Erlbaum, Mahwah, pp 269–299

Lenton T, van Oijen M (2002) Gaia as complex adaptive system. Phil Trans R Soc Lond B 357:683–695

Lovelock JE (1988) The ages of Gaia—a biography of our living Earth. The Commonwealth Fund Book program. Norton, NewYork

Newman M, Barabási A-L, Watts DJ (eds) (2006) The structure and dynamics of networks. Princeton University Press, Princeton/Woodstock

Newton PW (ed) (2008) Transitions, pathways towards sustainable urban development in Australia. CSIRO Publishing/Springer, Collingwood/Dordrecht

Olsson P (2011) Navigating transformations in social-ecological systems. NCCARF-seminar, Melbourne, 10 February 2011

Olsson P, Gunderson LH, Carpenter SR, Ryan P, Lebel L, Folke C, Holling CS (2006) Shooting the rapids: navigating transitions to adaptive governance of social-ecological systems. Ecol Soc 11(1):18. http://www.ecologyandsociety.org/vol11/iss1/art18/

Perez C (2002) Technological revolutions and financial capital. Edgar Elgar, Cheltenham

Peters J, Wetzels R (1997) Niets nieuws onder de zon en andere toevalligheden, Strategie uit Chaos. Uitgeverij Business Contact, Amsterdam/Antwerpen

Portugali J (2000) Self-organisation and the city. Springer, Berlin/Heidelberg/New York

Provincie Groningen (2000) Provinciaal Omgevingsplan, adopted 14 December 2000. Provincie Groningen, Groningen

Provincie Groningen (2006) Provinciaal Omgevingsplan, POP 2 text and maps, adopted 5 Juli 2006. Provincie Groningen, Groningen

Provincie Groningen (2009) Provinciaal Omgevingsplan 2009–2013, adopted 17 Juni 2009. Provincie Groningen, Groningen

Rip A, Kemp R (1996) Towards a theory of socio-technical change, Report prepared for Batelle Pacific Northwest Laboratories. Mimeao University of Twente, Enschede

Roberts K (2006) The lovemarks effect, winning the consumer revolution. powerHouse Books, New York

Roggema R (2007) Spatial impact of the adaptation to climate change in the province of Groningen, move with time. Climate Changes Spatial Planning & Province of Groningen, Groningen

Roggema R, van den Dobbelsteen A, Kabat P (2012) Towards a spatial planning framework for climate adaptation. SASBE 1(1):29–58

Rotmans J (2005) Duurzame samenleving: tussen droom en werkelijkheid staat complexiteit. Presentatie, 21 September 2005. Drift and Erasmus Universiteit, Rotterdam

Rotmans J, Kemp R, van Asselt M, Geels F, Verbong G, Molendijk K (2000) Transities en transitiemanagement: de casus van een emissiearme energievoorziening. ICIS, Maastricht

Scheffer M, Bascompte J, Brock WA, Brovkin V, Carpenter SR, Dakos V, Held H, van Nes EH, Rietkerk M, Sugihara G (2009) Early-warning signals for critical transitions. Nature 461:53–59

Schoot Uiterkamp T, van Dam F, Noorman K-J, Roggema R (2005) The Northern Netherlands: scanning the future. In: Van Dam F, Noorman K-J (eds) Grounds for change: bridging energy planning and spatial design strategies, Charette report. Grounds for Change/IGU, Groningen

Solé RV, Pastor-Satorras R, Smith E, Kepler TB (2002) A model of large-scale proteome evolution. Adv Comp Syst 5:43–54. In: Newman M, Barabási A-L, Watts DJ (eds) (2006) The structure and dynamics of networks. Princeton University Press, Princeton/Woodstock

Van den Dobbelsteen A, Roggema R, Stegenga K, Slabbers S (2006) Using the full potential – regional planning based on local potentials and exergy. In: Brebbia CA (ed) Management of natural resources, sustainable development and ecological issues. WIT Press, Southampton, pp 177–186

Vervoorn H (2003) Strategies of change. Presentation province of Groningen. A Room with a View, Amstelveen

Walker B, Holling CS, Carpenter SR, Kinzig A (2004) Resilience, adaptability and transformability in social–ecological systems. Ecol Soc 9(2):5. [online]: http://www.ecologyandsociety.org/vol9/iss2/art5

Watts DJ, Strogatz SH (1998) Collective dynamics of 'small-world' networks. Nature 393:440–442. In: Newman M, Barabási A-L, Watts DJ (eds) (2006) The structure and dynamics of networks. Princeton University Press, Princeton/Woodstock

Wilkinson C, Porter L, Colding J (2009) Metropolitan planning and resilience thinking: a practitioner's perspective. Crit Plan 17:24–45

Zuijderhoudt R (2007) Op zoek naar Synergie, omgaan met onoplosbare problemen. Academisch Proefschrift, UvA: Rob Zuijderhoudt Organisatieadviseur

Chapter 5
Networks as the Driving Force for Climate Design

Rob Roggema and Sven Stremke

Contents

5.1	Introduction	92
5.2	Network Theory	92
5.3	Explorations on Intensities	98
5.4	Application in the Peat Colonies	101
	5.4.1 Water Network	101
	5.4.2 Energy Network	102
	5.4.3 Transport Network	107
	5.4.4 Two Climate Designs for the Peat Colonies	108
5.5	Conclusion	114
References		115
Websites		115

Abstract In this chapter the potential transformation of an area and the role networks can play is discussed. For a far-future transformation, the current situation as well as the near-future, already taken policy decisions, function as the starting point for the design. Network theory is subsequently used to identify the crucial nodes in the networks where a potential transformation is likely to be successful. These nodes can be defined making use of the common rules of networks. Some points in networks are better (more intensively) connected with more links, than others. These hubs, the more attractive nodes to link with, get richer, which makes them even more attractive to link with, which makes them richer and so forth. The places where

---

R. Roggema (✉)
The Swinburne Institute for Social Research, Swinburne
University of Technology, PO Box 218, Hawthorn, VIC 3122, Australia
e-mail: rob@cittaideale.eu

S. Stremke
Landscape Architecture, Wageningen University and Research Centre,
PO Box 47, 6700 AA Wageningen, The Netherlands
e-mail: sven.stremke@wur.nl

these successful nodes are located can be identified and calculated as has been shown in the exercise in this chapter. The number and importance of connections as well as the typology of the nodes (a place consisting of one type is less attractive than if three networks overlap) play an important role in determining the interesting locations. Once these are found they can be used in the design, as is illustrated in the Peat Colonies case study. The structure of networks, with spines, nerves and nodes, in combination with a clear and specific objective leads to challenging and sustainable designs.

Keywords Network theory • Nodes • Intensity • Climate adaptation and mitigation • Spatial design

5.1 Introduction

When a region needs to undergo a transformation and include more climate adaptive design measures and strategies, current planning frameworks are not sufficient, as has been outlined in Chap. 4. One of the most crucial elements to influence in the spatial system, are the networks. Changes in the network types, structures and intensities determine changes in the system as a whole. Three time-horizons can be distinguished: now, near-future and far-future. 'Now' is reflecting the current situation. Networks are taken as unchanged in thinking about future changes. 'Near-future' includes all kind of policy decisions and plans that are already taken. A map of the near-future situation includes the networks as if they were already realised and taken is the base for future change. 'Far-future' identifies the potentially optimal and beneficial nodes and connections as well as required adjustments in current or near-future networks. Networks, as well as the focal points, the most important nodes are seen as a crucial basis for planning. Following the layer approach they form the first pair of layers and determine the major spatial directions for the longer-term future.

In this chapter theories about networks are briefly discussed and used to identify spatially the most important nodes and connections in networks. Subsequently this information is used to form the basis of a spatial planning framework for climate adaptive planning.

5.2 Network Theory

In this section theories about networks are briefly discussed and the concepts that are relevant and useful in spatial planning are illuminated.

Network theory is a field of computer science and network sciences and is also part of graph theory (the study of graphs and mathematical structures). Network theory is often deployed to examine the method of characterizing and modelling complex networks. Many complex networks share some common features, such as scale-free degree distribution. Network theory is applied in multiple disciplines, including biology,

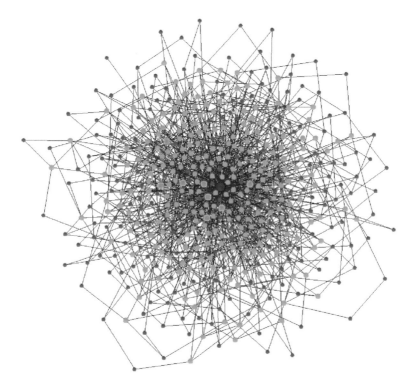

Fig. 5.1 Diagram of a scale-free network containing components with a highly diverse level of connectivity. Some components form highly interconnected hubs, while other components have few connections, and there are many levels of interconnectivity in between (Oikonomou and Cluzel 2006)

computer science, business, economics, particle physics, operations research and, most commonly, in sociology [www.techopedia.com/definition/25064/network-theory]. The interesting thing is that no matter for which discipline networks are used and analysed, the rules and laws are valid. The network architecture of evolution, the 'scale-free network' (Fig. 5.1), characterises the interaction network of proteins in yeast, worms, fruit flies and viruses, but also pervades social networks and computer networks, affecting, for example, the functioning of the World Wide Web (Oikonomou and Cluzel 2006).

All share common characteristics and many of the images showing striking similarities, such as social and self-organising networks (Fig. 5.2) the computer and economic networks (Fig. 5.3), the biological and neural networks (Fig. 5.4), the World Wide Web representations (Fig. 5.5) and the networks used in the Afghan stability operation (Fig. 5.6) and for the human genes (Fig. 5.7). In all the different networks some nodes are more crucial and have more connections than others. These are the nodes where major changes are more likely to take off. This information can be used to inform spatial planning. In Newman et al. (2006), the following network concepts are distinguished:

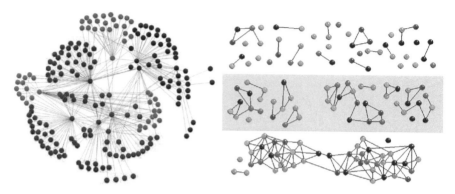

Fig. 5.2 Representation of a social network, *left* [http://en.wikipedia.org/wiki/Social_network, Copyrighted: Creative Commons] and self-organisation in networks, *right* (Nagler et al. 2011)

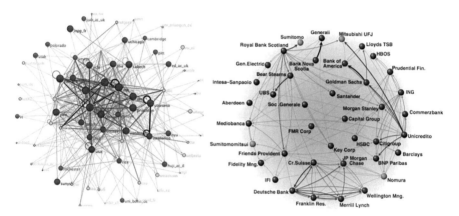

Fig. 5.3 Similarities in representation of a computer network, *left* [www.flickr.com/photos/anikarenina/238385060/] and an Economic network, *right* [www.mcn.ece.ufl.edu/public/YuejiaHe/network.htm]

1. For many years network theory was dominated by the belief that only random networks exist. The number of individual entities, nodes, was so large that whatever connections would be made, an even and equally distributed network would be the result. Only by adding edges, e.g. more contact points and therefore connections, the network, suddenly, gains quality. This theory of random network was developed and elaborated in several publications (amongst others: Erdós and Rényi 1960; Barabási 2003).
2. Later, dominant theory described the scale free network, in which certain nodes determine cores and other parts of the network then become periphery. This type of network can be described at any scale. The small world effect (Watts and Strogatz 1998) describes the characteristics of these networks: if the number of nodes in the network increases, while connected by a short path, which can be randomly added, the total length of paths will increase logarithmically and a high level of clustering will occur. These clusters are the core-groups in the network, connected by 'bridges' (Buchanan 2002), or the hubs and connectors (Barabási 2003);

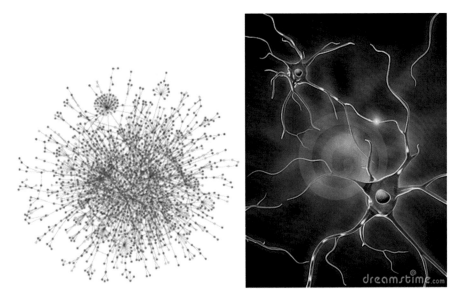

Fig. 5.4 The Biological network [www.thp.uni-koeln.de/~lassig/projects.html] and the Neural network [www.dreamstime.com/royalty-free-stock-photography-neural-network-image11819917]

3. The distribution of small and large hubs is described through a power law: there are only few (big) nodes in the network with many links and there are many small nodes with only few links (Barabási 2003; Buchanan 2002). Nodes that already have many links are more attractive to link with than small nodes, and the result of this is that rich nodes become richer;
4. The connectivity between nodes is increased through randomly additions of shortcuts and connections. The seemingly random new links will connect the nodes that are fittest in competing for links (Bianconi and Barabási 2001). This fitter-gets-richer (Barabási 2003; Buchanan 2002) phenomenon helps to understand the evolution of competitive systems in nature and society;
5. Robust networks are formed by interconnected nodes, which are highly clustered and know a minimum distance between any pair of randomly chosen nodes (Solé et al. 2002). This 'topological' robustness is rooted in the structural unevenness of scale free networks. The chance that a failure hits one of the few highly connected nodes in the middle of an endless amount of small nodes is minimal, which allows the network to recover and to keep functioning (Barabási 2003; Buchanan 2002);
6. Directed networks, such as for instance the World Wide Web, consist of a core (a giant strongly connected component), links-in and links-out as well as other islands and tendrils, represented visually by Broder et al. (2000) as a bow-tie (Fig. 5.8) (Barabási 2003);

When these network characteristics are translated to spatial structures and elements, as demonstrated by Castells (1996) or Graham and Marvin in their Splintering

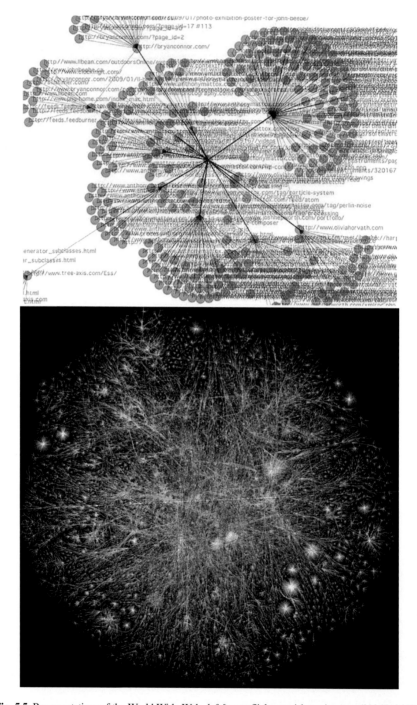

Fig. 5.5 Representations of the World Wide Web, *left* [www.flickr.com/photos/amattox/3236510649/], and *right* [www.mirror.co.uk/news/uk-news/20th-anniversary-of-world-wide-web-381974, Copyrighted: http://creativecommons.org/]

5 Networks as the Driving Force for Climate Design

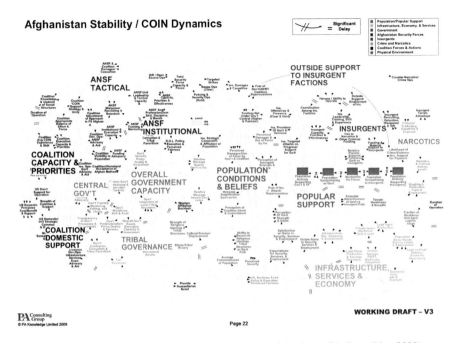

Fig. 5.6 Strategic network for the Stability operation in Afghanistan (PA Consulting 2009)

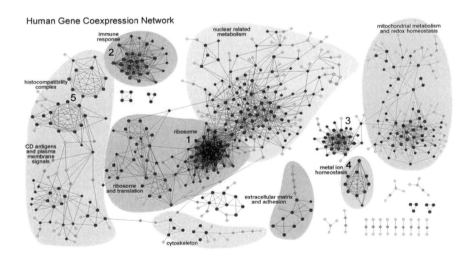

Fig. 5.7 Graphical view of the Human gene co-expression network, where the nodes correspond to genes and the edges to co-expression links (Prieto et al. undated)

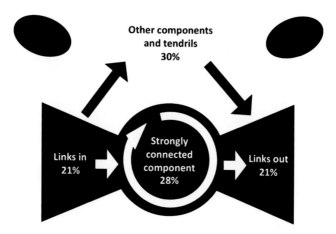

Fig. 5.8 The 'Bow-tie', with a core, links and islands and tendrils (Broder et al. 2000)

Urbanism theory (2001) of super-positioning intertwining infrastructure,[1] the following general rules of network theory made useful for spatial planning were found:

- The amount of nodes in combination with certain random connections (creating a 'small world');
- Presence of 'rich' cores with many connections;
- Existence of 'fit' nodes, attractive to connect with;
- Clustering and minimal distances in combination with peripheric areas where small nodes with few links and longer distances are found;
- 'Islanding' of several parts only accessible for specific functions, connected through a giant strongly connected core.

These principles determine specific spatially defined locations within networks, where the most likely changes might take place. When an area needs to transform to a more resilient region, these starting points for change or, where novelties can be developed that offer the change of changing the current stable regime, need to be identified.

5.3 Explorations on Intensities

The application of the above principles is in essence an exercise to discover the intensity and importance of nodes in the network. The places where a clustering of strong and important nodes occurs are seen as the richest and fittest nodes. Where

---

[1] Graham and Marvin (2001) describe the intertwined infrastructures of electropolis (energy), hydropolis (water), cybercity (Internet), railcity (train) and autocity (car) super-positioned on top of each other.

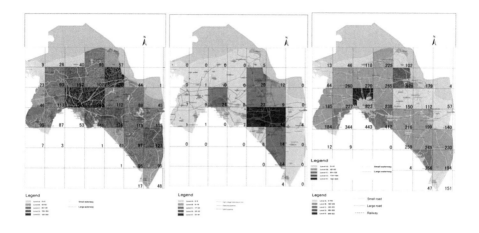

Fig. 5.9 The network maps of water, energy and transport (Hao and Wang 2010)

many nodes are present within a small area the connectivity is higher. This intensity represents the richness of the node. Where nodes of many different functional types, such as water, energy or transport, exist, the more attractive it is to link. The combination of the two reflects an overall value for an area. The higher the value the more likely new links will develop, adding strength and growth to the existing cluster.

This exercise has been conducted in Groningen province (Hao and Wang 2010). The networks were analysed in three steps: (1) the density of individual nodes per area and (2) the number of different overlapping network types at one physical location. The final step (3) in this exercise was the combination of the two steps in one overall value per grid-cell.

1. The first step examined the density of nodes, defined as: the number of nodes within a grid-cell of 10 × 10 km, combined with their importance (the bigger, the more important). In this study the importance was weighted as follows: a minor infrastructural element, such as a small local road, a little stream or a household electricity gridline counts for a factor 1, a mediocre element counts for a factor 3 and a major element, such as a freeway, high voltage power-line or a main canal or river counts for a factor 5. The number of nodes times their respective importance gives the value for each grid-cell. The results, in the form of specific maps for energy, water and transport, are shown in Fig. 5.9, in which the darkest colours represent the most intense places;
2. The second step analysed the number of different network types that form a node: single (only water, energy or transport), double (overlap of any combination of two out of three) or triple (all of them overlap) nodes. In case of a double node, the calculation is multiplied by a factor 10 and in case of triple node by a factor 100. Within each grid cell of 10 × 10 km, each node was multiplied with the appropriate factor and the total value was calculated by adding all node values to reach a total score per cell;

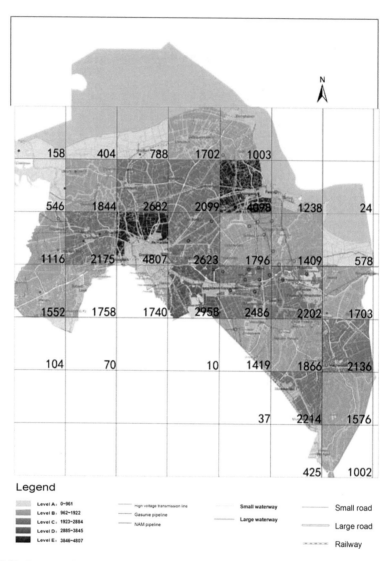

Fig. 5.10 Integrated intensity of networks for Groningen area (Hao and Wang 2010)

3. In step three, the values of first step, individual networks and second step, network types were added to give the total score for each grid cell (Fig. 5.10). The darkest colour represents the highest values. Higher values imply stronger and more intense, well-connected clusters, where change is more likely to occur.

The integrated map illustrates that in certain grid-cells change is more likely to start than in others. When an area needs to undergo change, these highly intense and well-connected clusters offer the highest probability.

5.4 Application in the Peat Colonies

A good example of how network analysis informs climate design is the Peat Colony area, where both adaptation and mitigation issues have been integrated in the design. The current Peat Colony area is typically characterized through large agricultural fields. Almost 60% of the area is arable land, followed by 14% cattle grazing. In contrast with the Hondsrug ('Dog ridge') in the west, the Peat Colonies contain only a small forest area (7%) and heather (2%). Considerable surface is devoted to industry (1,451 ha) and greenhouses (288 ha). With regard to network elements, the Peat Colonies have a dense and extensive network of small ditches (total of 1,974 km) and large ditches (656 km). The transportation system consists of streets (1,903 km), local roads (960 km), regional roads (378 km) and main roads (247 km). Currently, there is only 12 km of highways present. A total of 154 km of high voltage electricity lines are found in the region.

In order to envision alternatives futures for the long-term development of the Peat Colonies, it is not only important to map current conditions in the area but also necessary to have a good understanding of possible development in the near future (Fig. 5.11). Development of nature areas presents the largest possible land-use change in the near future.

5.4.1 Water Network

The Peat Colonies are known for their straight canal structure (Fig. 5.12), which was created to dry the area and transport peat to cities elsewhere in the country. Ditches less than 3 m wide (1.974 km) and canals between 3 and 6 m wide (656 km) constitute a fine grid of water network, which again are connected via the larger canals (189 km) (Fig. 5.12). The total surface of open water in the region amounts to 2.892 ha, which is about 4% of the total surface.

The water network is very dense and consists of numerous small and tiny ditches, especially in the North Western and central parts of the Peat Colonies. Several larger waterways cross the area from South to the North, of which the Hunze (the most western one), the Mussel Aa and Ruiten Aa (both in the East) are natural formed rivers. The major artificial canals, the Stads Canal and the Wildervanks Canal run from South East to the North. There is a string of waternodes found in the southern area, connected to the Verlengde Hoogeveense Vaart as well as between the two parallel running canals of Stadskanaal and the Wildervanks Canal all the way from Ter Apel in the South via Stadskanaal towards Veendam. A same string of potential rich nodes is identified in the parallel running Ruiten Aa and Ruiten Aa Canal to the East, but outside the study area. These (potentially) intense nodes (Fig. 5.13) form the points in the water network, which are most likely to be developed. In case new canals will be created, the nodes change accordingly.

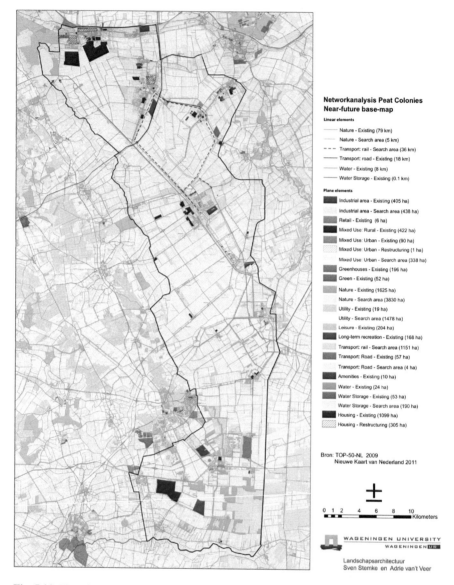

Fig. 5.11 Near-future base map Peat Colonies [www.kaart.nieuwekaart.nl]

5.4.2 Energy Network

The energy network (Fig. 5.14) of the Peat Colonies consists of high-voltage electricity lines (154 km), gas networks (national and NAM: 123 km) and regional and local gas pipelines (350 km), an old oil pipeline and at least two heat networks. The energy demand is for the largest part determined by industrial uses and built-up areas.

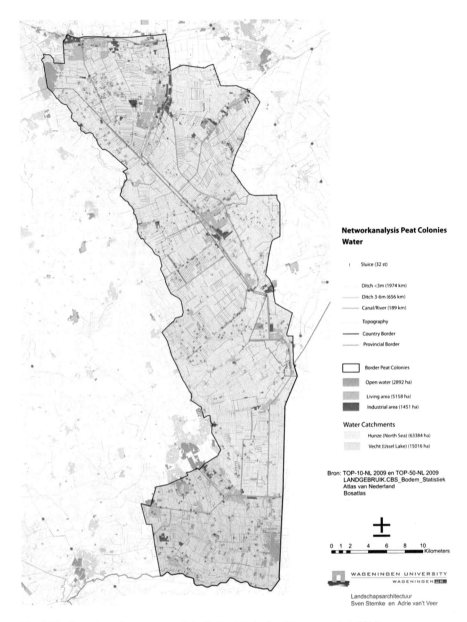

Fig. 5.12 Overview of water network in the Peat Colonies (Broersma et al. 2011)

The gas distribution network is very dense in built-up areas, but limited to the national grid outside these areas. The main gas-line crosses the area from North, where the gas is extracted, to the South. This gas-line crosses the area without having any relation within the area at the moment. There are potentially places where nodes of energy exchange can be created, for instance where the gas-line crosses main transport infrastructure, such as the provincial roads. These connection points form

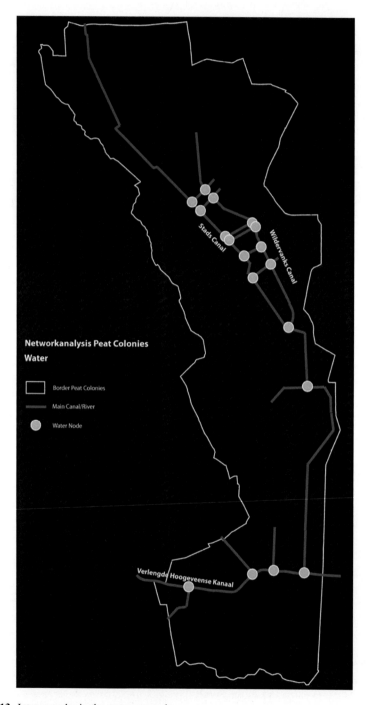

Fig. 5.13 Intense nodes in the water network

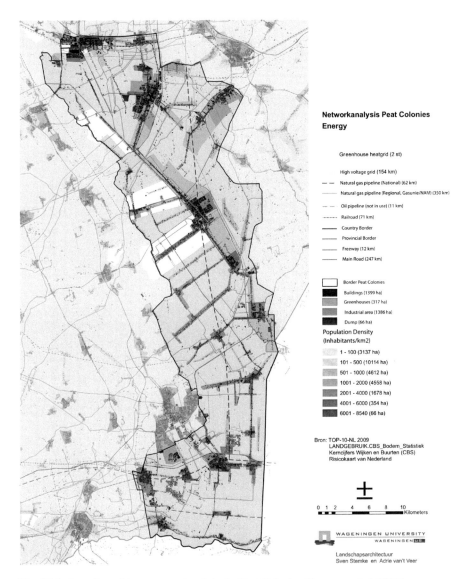

Fig. 5.14 Overview of the energy network in the Peat Colonies (Broersma et al. 2011)

nodes of intensity (Fig. 5.15), where new urban developments are more likely to start, given the presence of energy and access. Other potential development areas, are found where the gas-line and existing urban areas meet, such as near Emmen and Stadskanaal. The high-voltage line flows a slightly different route, but also crosses from North to South. For this line the same is true as for the gas-line; it hasn't many relations with the area yet. These might also be developed, for instance where this line crosses the main transport network, crosses the gas-line or can be directly linked with existing built-up areas, such as Veendam, Hoogezand, the Pekela's, Musselkanaal, Ter Apel and Emmen.

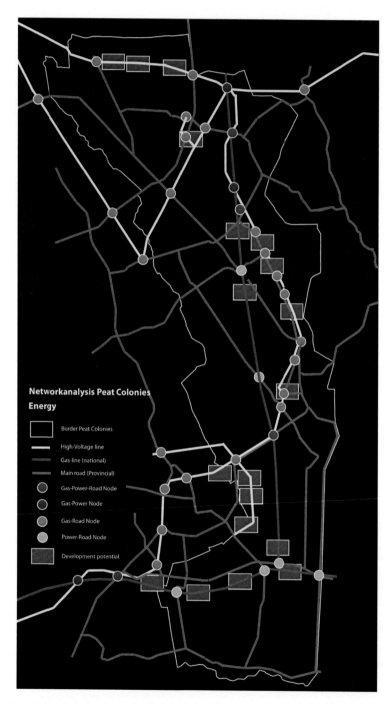

Fig. 5.15 Nodes in the energy network

5.4.3 Transport Network

The transportation network (Fig. 5.16) is similar to many other border regions and is rather underdeveloped. There exists an extensive road network (3.487 km). Railroads from the North and from the South are disconnected. An extensive bus transport network runs through the area and connects the main settlements.

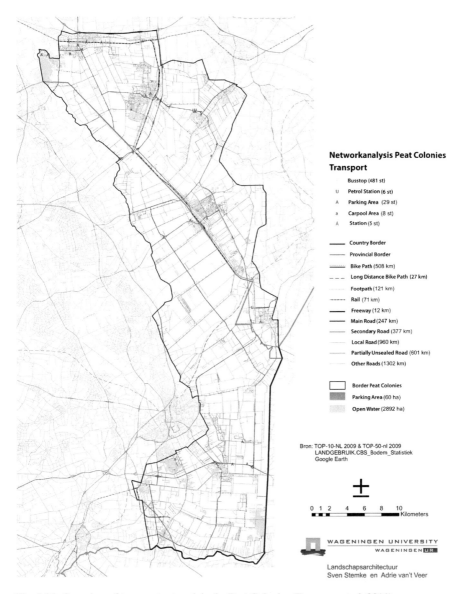

Fig. 5.16 Overview of transport network in the Peat Colonies (Broersma et al. 2011)

The main elements in the transport network are the main roads inside the area. Freeways are only present in the South and just outside the study area, North of Hoogezand. The main roads are of provincial importance and connect Ter Apel-Groningen and Emmen with Veendam. Both routes are crossing each other near Wildervank, which makes this a potential intense node in the transport network (Fig. 5.17). Where long-distance bike roads and railroads cross the car network, potential intense nodes are identified. These nodes and the existing petrol stations and railway stations complete the field of important nodes. Main infrastructure adjacent to existing built up areas, is potentially likely for emergent urban developments, in this case the most obvious, such as Emmen, Stadskanaal, Ter Apel, Veendam and Hoogezand.

5.4.4 Two Climate Designs for the Peat Colonies

On the basis of the network analysis of the water, energy and transport network two spatial models have been designed: 'Lonelycolony' and 'Peatcometro' (Broersma et al. 2011). The places where interventions are proposed were, in both models, determined through the location of the strongest, best-connected (the 'richest and fittest') clusters of nodes. Each of the climate designs take a sustainable energy supply as the starting point, but integrate climate adaptation strategies and measures in the designs.

5.4.4.1 Lonelycolony

The 'Lonelycolony' model (Fig. 5.18) is based on the aim to solve the energy demand on a scale as local as possible. Therefore, small-scale, decentralised renewable energy supply is proposed, which needs to lead to self-sufficiency within individual municipalities. All locally available energy potentials will be used to save energy and supply from renewable resources. As a result, in this model the Peat Colonies will become autonomous, e.g. it becomes independent from import and variable, rising oil prices. Because of the fact that many strategies and measures are found at the local scale, it improves opportunities for job creation and stimulates the regional economic development. Moreover, the short transportation distances imply minimal energy loss.

The network analyses on water, energy and transport are used to design this model. The main choice has been to create strong clusters in core settlements in each of the municipalities. The crucial clustering is located where the central spine in the water network, the main canal with its ramifications, links with the transportation network. The places where locally available energy potentials make a self-sufficient energy supply possible form the core developmental areas. In each of the municipalities one or two of these small-scale places the heat supply (rest-heat from industry, geothermal heat and greenhouses) is provided through underground networks. Each individual house is stimulated to generate its own energy, but can use centrally generated electricity

5 Networks as the Driving Force for Climate Design

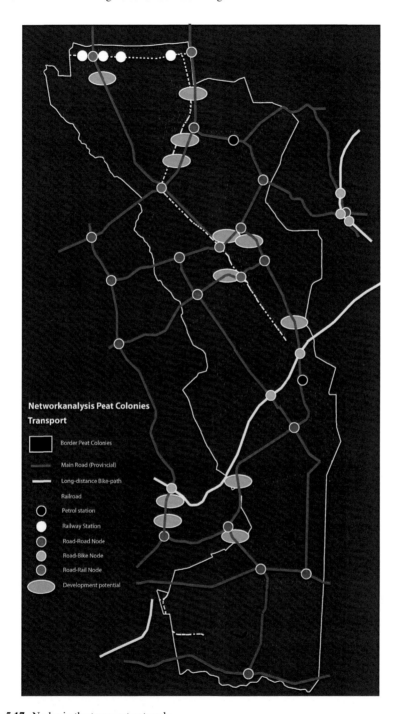

Fig. 5.17 Nodes in the transport network

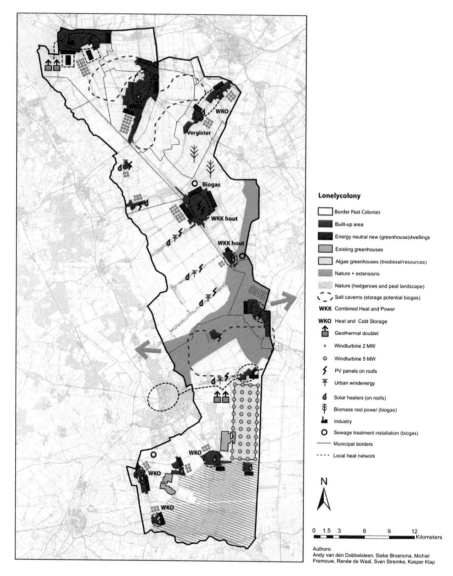

Fig. 5.18 The climate design 'Lonelycolony' (Broersma et al. 2011)

from small wind-turbines, of which several are placed in each core. New houses can be built under the condition that they will be made energy-neutral. Outside these core areas, households need to provide their own electricity and heat supply (PV, solar heating and small-scale wind-power). In addition, centrally located extensions of existing greenhouse complexes and a wind-park supply the residual demanded heat and electricity.

The most intense places in the water network determine the location for an area, where additional water-storage, nature- and forest area is developed. From this area

5 Networks as the Driving Force for Climate Design

the surplus of biomass, is used to produce heat and electricity using bio-CHP's (biomass Combined Heat and Power installations).

5.4.4.2 Peatcometro

The 'Peatcometro' model (Fig. 5.19) uses the available renewable resources efficiently and on a large-scale. On top of the ambition to become fully self-sufficient in the Peat

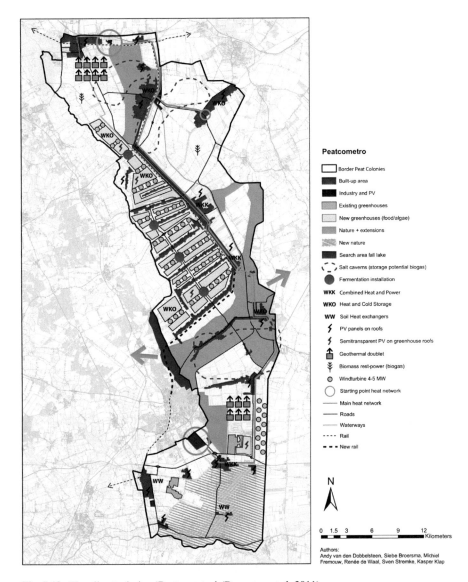

Fig. 5.19 The climate design 'Peatcometro' (Broersma et al. 2011)

Colonies, the potential to become a net exporter of energy is explored in this model. All local potentials to supply energy are fully used in order to provide the area itself and the surrounding, more urbanized, areas. Renewable resources are used to generate energy in a centralised way, occupying a large production landscape to maximise the supply of sustainable energy, such as large-scale algae-breeding, lairaged in innovative greenhouses, delivering bio-diesel. In between the green houses large-scale wind-turbines are realised producing, in combination with semi-transparent PV-foil a large amount of electricity. Both the placing of wind-turbines and the PV-foil on green-house roofs mean multiple use of space, a very efficient and intensive way of energy production. The rest-heat from the greenhouses, together with geothermal heat, is used to heat houses and for export to surrounding areas.

The networks of transport, energy and to a lesser extent water shape the Peatcometro design. Along the central spine of transport routes and its ramifications the exchange of energy is organised, allowing the production units, the greenhouses, which are positioned in between the side nerves of the system (consisting of existing water network, existing roads and additional energy transportation infrastructure), to connect easily to the network. The water in these side-nerves of the networks is essential to supply the algae production with enough, constantly available and clean water. The location of those side waterways determines therefore the structure of the climate design. At the end (or beginning) of these side nerves, new nodes and hubs are introduced where energy from the distributed production units links with the central exchange system. At several places the major networks connect with the surrounding areas in order to make further distribution possible. The major spines in the water network form the base to develop huge forest and nature areas.

5.4.4.3 Integrated Design

Both models were subsequently integrated in one renewable energy vision (Fig. 5.20), which formed the basis for a structure image, in which all ingredients were represented. The existing water network is taken as the basis for the structure image. The spatially open structure of canals, the 'wijken' (smaller side-canals), often refilled with water, and ribbons (linear villages) is reinforced. Within this main structure two low-lying areas are reserved for floating algae greenhouses, in combination with storage of surpluses of rainwater. These areas function as the connecting zones of 'waterfarms' between the eastern and western parts of the Peat Colonies. Economical, ecological, energetic and water functions are combined in these additions to both the water and energy network. The best-connected nodes in the energy network are used to form the starting points of the heat networks. These isolated areas function in the beginning as solitary elements, but can be connected with each other using the energy network, at a later stage. One big robust heat network emerges. Self-sufficient villages are combined with decentralised large-scale energy generation in the South East. Here the innovative Algae greenhouses

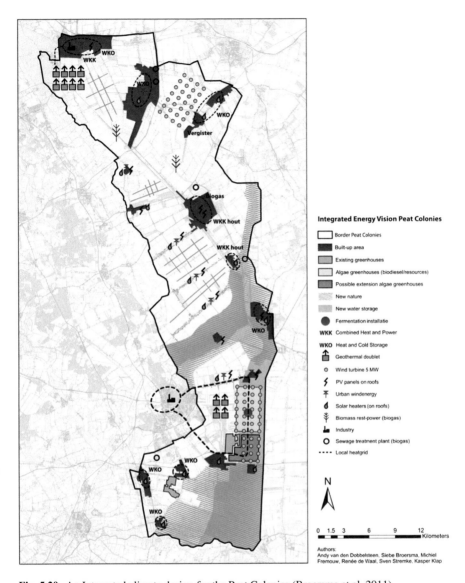

Fig. 5.20 An Integrated climate design for the Peat Colonies (Broersma et al. 2011)

are projected as well. A heat-ring is projected to connect these greenhouses with geothermal production areas and transport the heat towards consumers in the larger towns. Other towns and villages are provided with heat from local supplied renewable energy sources, of which the location is based on the local available renewable energy potentials of heat-generation and storage in the soil.

5.5 Conclusion

Network theory offers a wide range of insight, which can be made useful for spatial planning. Especially when a transformation in a certain area is required, and it often is in case of climate change, the rules and laws of networks are helpful to identify the places in the area where the transformation is likely to be started.

The intensity, richness and fitness of nodes in networks determine to a high extent the importance of these nodes. The more important they are, the more clustering takes place and the more links with other nodes are developed. This makes these intense clustered nodes the places where it is highly likely that new developments or innovation will kick off. In a spatial sense the places where these nodes are located, form interesting places to look for possible change and the start of transformation.

The exercise to identify these places in the Dutch province of Groningen can be seen as the successful first step in identifying the areas where these intense nodes can be found. The case illustrates for the water-, energy- and transport-network that both intensity (richness) as importance and attractiveness (fitness) can be determined and located. These results can subsequently be used in spatial planning and design to locate core areas for intense mixed use and combination of functions.

In the extensive example of the Peat Colonies the detailed network analyses of the three network of water, energy and transport forms the basis for two different climate designs. In these designs climate adaptation measures are combined with mitigation measures, mainly in the energy domain. The results show that the network analyses can be very well used to inform the designs and that, based on the same analysis two fundamentally different models can be designed, both meeting the requirements to become energy neutral and combine adaptation and mitigation measures. In the integrated design however, the strength of the models is somewhat decreased, due to necessary combination of measures originated from two fundamentally different models as well as the influence of 'regular policy', which pursued to support former policies. Despite the fact that the actual design problem was to design a region, which could become a net carbon-sink, a structural different assignment than ever before, recent policies, meant to deal with problems of the past, were pushed to solve the new problem. Which proves to be unsuccessful. It may be concluded here that, when a structural new type problem appears, the old policies need to be excluded from the process.

The Peat Colony example shows that taking a network based approach the major, well-connected nodes in the diverse network can be used in different ways to base the design work on. The spines, nerves and nodes of the networks, in combination with clear objectives ('the adaptive area as a carbon sink'), are easily useable in the design process and are capable of shaping in diverse ways interesting and sustainable designs.

References

Barabási A-L (2003) Linked: How everything is connected to everything else and what it means. New York: Penguin Group

Bianconi G, Barabási A-L (2001) Competition and multiscaling in evolving networks. Europhys Lett 54:436–442. In: Newman M, Barabási A-L, Watts DJ (eds) (2006) The structure and dynamics of networks. Princeton University Press, Princeton/Woodstock

Broder A, Kumar R, Maghoul F, Raghavan P, Rajagopalan S, Stata R, Tomkins A, Wiener J (2000) Graph structure in the web. Comput Netw 33:309–320. In: Newman M, Barabási A-L, Watts DJ (eds) (2006) The structure and dynamics of networks. Princeton University Press, Princeton/Woodstock

Broersma S, Fremouw M, Stremke S, van den Dobbelsteen A, de Waal R (2011) Duurzame energiestructuurvisie voor de Veenkoloniën. TU Delft/WUR, Delft

Buchanan M (2002) Nexus, small worlds and the groundbreaking science of networks. W.W. Norton & Company, Inc., New York

Castells M (1996) The rise of the network society. Blackwell, Oxford

Erdős P, Rényi A (1960) On the evolution of random graphs. Publ Math Inst Hung Acad Sci 5:17–61. In: Newman M, Barabási A-L, Watts DJ (eds) (2006) The structure and dynamics of networks. Princeton University Press, Princeton/Woodstock

Graham S, Marvin S (2001) Splintering urbanism: networked infrastructures, technological mobilities and the urban condition. Routledge, London

Hao H, Wang X (2010) Spatial planning for climate change adaptation – test and improve spatial planning methodology for the CCA. Thesis. Van Hall Larenstein, Velp

Nagler J, Levina A, Timme M (2011) Impact of single links in competitive percolation. Nat Phys 7:265–270

Newman M, Barabási A-L, Watts DJ (eds) (2006) The structure and dynamics of networks. Princeton University Press, Princeton/Woodstock

Oikonomou P, Cluzel P (2006) Effects of topology on network evolution. Nat Phys 2(8):532–536

PA Consulting (2009) Dynamic planning for COIN in Afghanistan. http://msnbcmedia.msn.com/i/MSNBC/Components/Photo/_new/Afghanistan_Dynamic_Planning.pdf

Prieto C, Risueno A, Fontanillo C, de las Rivas J (undated) Human gene coexpression landscape: confident network derived from tissue transcriptomic profiles. Bioinformatics and Functional Genomics Research Group, Cancer Research Center (CIC-IBMCC, USAL/CSIC), Salamanca. Retrieved from: http://bioinfow.dep.usal.es/coexpression/. Accessed 9 Feb 2012

Solé RV, Pastor-Satorras R, Smith E, Kepler TB (2002) A model of large-scale proteome evolution. Adv Complex Syst 5:43–54. In: Newman M, Barabási A-L, Watts DJ (eds) (2006) The structure and dynamics of networks. Princeton University Press, Princeton/Woodstock

Watts DJ, Strogatz SH (1998) Collective dynamics of 'small-world' networks. Nature 393:440–442. In: Newman M, Barabási A-L, Watts DJ (eds) (2006) The structure and dynamics of networks. Princeton University Press, Princeton/Woodstock

Websites

http://en.wikipedia.org/wiki/Social_network. Accessed 9 Feb 2012

www.dreamstime.com/royalty-free-stock-photography-neural-network-image11819917. Accessed 9 Feb 2012

www.flickr.com/photos/amattox/3236510649/. Accessed 9 Feb 2012

www.flickr.com/photos/anikarenina/238385060/. Accessed 9 Feb 2012

www.mcn.ece.ufl.edu/public/YuejiaHe/network.htm. Accessed 9 Feb 2012
www.mirror.co.uk/news/uk-news/20th-anniversary-of-world-wide-web-381974. Accessed 9 Feb 2012
www.kaart.nieuwekaart.nl. Accessed 15 Nov 2011
www.techopedia.com/definition/25064/network-theory. Accessed 9 Feb 2012
www.thp.uni-koeln.de/~lassig/projects.html. Accessed 9 Feb 2012

Chapter 6
Swarm Planning Theory

Rob Roggema

Contents

6.1	Introduction	118
6.2	Problem Statement	119
6.3	Approach	119
6.4	Current Planning Paradigms	120
	6.4.1 A Selection of Prevailing Planning Paradigms	120
	6.4.2 A Review of 2 Years of Planning Journals	124
6.5	Exploring Complexity	126
	6.5.1 Complexity Theory	126
	6.5.2 Cities as Complex Systems	129
	6.5.3 Use of Complexity in Planning	130
	6.5.4 Proposition: Swarm Planning	131
6.6	Conclusion	133
References		134

Abstract In this chapter a new planning theory is developed. The rationale for needing this new theory lies in the fact that current spatial planning paradigms both seen from an academic as practice perspective, lack the possibility to deal with problems that are not straightforward, clearly defined and predictable: wicked problems. The majority of planning literature is still focusing on well-known problems and is operational within a governmental context. Despite the fact that a debate is emerging about the need for planning approaches that incorporate dynamic environments, look at the future from a change perspective and focus on the emergence of spatial order

R. Roggema (✉)
The Swinburne Institute for Social Research, Swinburne
University of Technology, PO Box 218, Hawthorn, VIC 3122, Australia
e-mail: rob@cittaideale.eu

initiated by key actors outside government, recent publications show that 94% of the articles discuss traditional topics and approaches.

If planning needs to be prepared to incorporate wicked problems it is attractive to use complexity theory, which deals with complex adaptive systems. However, the majority of research in complexity theory in relation to planning and cities focuses on the understanding of emergence and self-organisation by developing ever more advanced computational models. This mathematicalisation of the city distracts the attention from intervening in these systems to improve preparedness in dealing with wicked problems.

The gap as shown above can be filled through the launch of a planning theory that deals with unpredictability of the future and incorporates complex systems behaviour. The theory is called Swarm Planning, because it emphasises swarm behaviour of the system to be beneficial for the overall resilience and lessen the impact of uncertainties, complexity and change.

Swarm Planning introduces two planning strategies: intervention in the system as a whole and free emergence through the attribution of individual components with Complex Adaptive System (CAS)-properties in order to perform self-organisation.

Keywords Planning theory • Swarm Planning • Complexity • Climate change • Wicked problem

6.1 Introduction

Climate change adaptation is seen as a wicked (VROM-raad 2007; Commonwealth of Australia 2007) or even a superwicked (Lazarus 2009) problem. A wicked problem is accurately defined in the seminal paper of Rittel and Webber: "Dilemmas in a General Theory of Planning" (Rittel and Webber 1973). Wicked problems are defined as being dynamic, they do not know a final solution, are "a one shot operation" and essentially unique. As planners, we do not have the right to be wrong.

Spatial planning is defined in many different ways. Dror for example (1973) describes planning as a process: "Planning is the process of preparing a set of decisions for action in the future, directed at achieving goals by preferable means". In the course of this paper spatial planning is defined as the 'co-ordination, making and mediation of space' (Gunder and Hillier 2009: 4).

Current (and historic) discourses in spatial planning, such as incrementalism (referring to Lindblom 1959), post-positivism (as described in Allmendinger 2002a), communicative planning (amongst others: Habermas 1987, 1993; Healey 1997; Innes 2004), agonism (see: Mouffe 1993, 2005; Hillier 2003; Pløger 2004), reflexive planning (Beck et al. 2003; Lissandrello and Grin 2011) or even the actor network approach (Boelens 2010) do have considerable difficulties to deal with wicked problems, or solutions, or fail to take wicked problems as the subject of planning. Hence, the need for an alternative theory emerges. In this paper this theory, Swarm Planning, is explored and developed.

6.2 Problem Statement

Our world becomes increasingly complex and turbulent (see for instance Ramirez et al. 2008), as reflected in the fields of energy (peak oil and consequences of oil prices (Campbell and Laherrrère 1998; Campbell 1999, 2002a, b; Rifkin 2002; Belin 2008; Sergeev et al. 2009)), accelerated climate change (Tin 2008; Richardson et al. 2009; PBL et al. 2009; Sommerkorn and Hassol 2009), but also in the global economy. More specifically, climate adaptation, defined as a wicked problem itself, but also energy systems planning (Van Dam and Noorman 2005) are marginally connected with the spatial planning domain. This means that, inevitable, adaptation and energy planning take place in separate world, where they actually are not 'planned' as spatial systems. Meanwhile, regular planning (e.g. (urban) developments) continue to take place.

These problems are wicked and spatial planning lacks the processes, decision-making and tools to uptake them. Thus the problem can be stated as:

- Spatial planning is not used as a platform or framework for 'solving' these problems;
- Current spatial planning paradigms themselves predominantly focus on decision making within government for a well-described (planning) problem. Within planning theory there is a lack of methods and planning approaches for wicked problems.

While changes increasingly appear in a non-linear fashion, spatial planning increasingly lacks answers.

6.3 Approach

The research presented in this chapter distinguishes several pieces of work (Fig. 6.1).

In Sect. 6.4 a literature review about current planning paradigms has been conducted in two different ways. In the first place (Sect. 6.4.1) current paradigms as well as from the past have been identified and analysed on their usefulness for wicked problems. Secondly (Sect. 6.4.2), articles, published in 2010 and 2011 in four international planning journals (Planning Theory, Planning Theory and Practice, Australian Planner and European Planning Studies), were analysed on the merits of containing theories useful to complex problems. This illuminates the common typology of current subjects in planning journals.

In Sect. 6.5 complexity (Sect. 6.5.1) and planning for cities (Sect. 6.5.2) has been explored. On the one hand side because cities or areas are seen as complex adaptive systems (Portugali 2000; Batty 2005; Allen 1996; Dos Santos and Partidário 2011), but on the other hand the insights from complexity theory could be of use to develop a planning approach capable of dealing with wicked problems. The central question (Sect. 6.5.3) has been if current planning paradigms and/or in combination with scholarly writing on complexity and planning are sufficient of being able to make plans for wicked problems? The answer to this question led to the development of

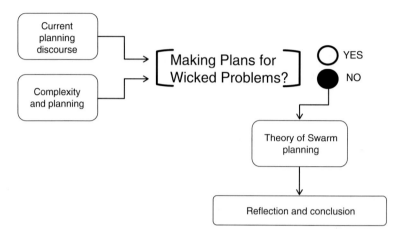

Fig. 6.1 Schematic representation of the research approach

Swarm Planning Theory (Sect. 6.5.4). The theory of swarms (Fisher 2009; Miller 2010) and tipping points (Gladwell 2000), lessons from complexity (amongst others: Schwank 1965), existing examples of creating plans using the understanding of swarms (Oosterhuis 2006, 2011) and the use of complex adaptive systems properties (Roggema et al. 2012a) all were used to develop Swarm Planning theory, which enables planning to incorporate wicked problems. Finally (Sect. 6.6), the developed theory has been critically reflected upon and conclusions were drawn.

6.4 Current Planning Paradigms

In this section a brief overview of spatial planning paradigms is presented. Despite the fact that it is hardly possible to do justice to existing planning theories and paradigms in one paragraph each, it attempts to capture the main characteristics in order to come to a judgement-*light* of the applicability of each to deal with wicked problems. In-depth study and elaboration is required to provide a more thorough basis for the judgements. The current planning paradigm is analysed in two ways. In the first section a selection of well-known paradigms will be briefly described and their eventual shortcomings in the face of dealing with wicked problems will be examined. The second section will look into all articles published in four international planning journals over the years 2010 and 2011.

6.4.1 *A Selection of Prevailing Planning Paradigms*

In recent planning literature sparse, but strong signals can be found illuminating a change in planning paradigm. Scholars such as Newman, Boelens, Miraftab, Davy and Gunder all, from different angles, point at (the need for) planning 'moving away' from its traditional base: the government. Past planning paradigms, such as positivism, incrementalism, post-positivism, agonism and reflexive planning are all to a certain extent, but mostly inextricably connected to governmental agencies. These paradigms described briefly and their applicability to plan for wicked problems is considered.

The first discourse is **positivism**, which is build on the belief that data informs how to plan, and the logic of large amounts of data leads us to the one and only truth. Positivist planning schools look for general laws, are science based and top down organised (Allmendinger 2002b). For example, comprehensive rationality aims to understand the 'whole' through a thorough analysis of everything before problems can be defined and solved. And in systems theory cities are seen as systems, which can be modelled and changes can be predicted, once all characteristics are known (McLoughlin 1969). Positivist approaches require much data before conclusions can be drawn or plans can be made. For many organisations the collection and analyses, even with advanced computers, are hardly possible (Banfield 1973) and require large investments.

> Especially when problems are complex, and most of the issues are nowadays (De Roo 2006), a rational comprehensive planning (positivist) approach is not possible and serious simplification of the problem is necessary (Lindblom 1959). Even though the increased computer capacity nowadays allows for dizzying calculations, it still is questionable whether the answers provided will deal with hardly predictable and surprising wicked problems. The question may be raised if, in order to deal with wicked problems, investments in deepening data collection and understanding the system are the most economic choice and, more in general, if a positivist approach will satisfy.

The second discourse, **incrementalism,** is extensively described in the seminal paper "the science of muddling through" (Lindblom 1959). Because it is not possible to comprehend all information, Lindblom elaborates on what he calls the method of Successive Limited.

Comparisons. It considers planning as a process of continually building out from the current situation, step-by-step and by small degrees. This incremental way is an adequate method for policy making if present policies are satisfactory, the nature of problems and the available means to deal with problems have a high degree of continuity (Dror 1964). However, incrementalism is an adversary process culminating in compromises of which the decisions tend to reflect the values of those in power, being a *status quo*. (Cates 1979).

> When the environment presents itself as a non-incremental change, as many wicked problems do, this approach doesn't work, because these changes are too large to respond to incrementally. In Lindbloms mindset: if the 'limited comparison' consists of two, of which one is a step change, the administrator would randomly choose the one implying limited change. Recent research illuminates climate change as a phenomenon characterised by 'step changes' (e.g. a significant change) (Jones 2010).

Under the umbrella of the **post-positivist** discourse (Allmendinger 2002a) several planning 'schools' share similar characteristics: a focus on subjective knowledge and endless possibilities for description (Allmendinger 2002b). Or, as Farmer wrote (1993: 392): post-structuralism could be characterised by its rejection of 'master narratives' and 'foundational claims that purport to be based on science, objectivity, neutrality' (cited in: Hillier and Cao 2011). Post-structuralism describes social and cultural systems (including cities) that are relational, open and dynamic, constantly in the process of emerging or 'becoming' different. Spaces and places are always in the process of being made and are unpredictable, especially in the longer term.

Examples of the, partially overlapping, post-positive planning 'schools' are collaborative planning (Healey 1997; Innes 2004; Innes and Booher 1999, 2004) or communicative rationality (Habermas 1987, 1993), post-modernism (Beauregard 1996; Jencks 1987; Allmendinger 2001) and communicative planning (Forester 1989).

In one of the central texts of this paradigm, Healey calls upon stakeholders to '… take a major leap in reflexive activity, to stand back from their particular concerns, to review their situation, to re-think problems and challenges, to work out opportunities and constraints, to think through courses of action which might be better than current practices and to commit themselves to changing things' (Healey 2006: 244).

In general, post-positivism is oriented on structuring processes, stakeholder involvement and aiming for consensus.

> Following Newman (2011): In assuming that communication and dialogue can operate in a neutral framework, collaborative planning theory imagines a level playing field where differences in power and wealth are somehow counteracted. Yet, we see how this formal neutrality and equality – where everyone is included as a 'stakeholder' – can function in an ideological way to legitimize an already assumed economic consensus, while de-legitimizing antagonism and dissent as irrational, violent and undemocratic.

In contrast with the consensus-oriented post-positivists, **agonism**, as Mouffe (1993, 1999, 2000, 2005), Hillier (2003) and Pløger (2004), amongst others are discussing, acknowledges and respects permanent conflicts in political communication. According to Mouffe, "the aim must be to transform an 'antagonism' into 'agonism' between 'adversaries' rather than 'enemies'. In the political realm of agonism, compromises and consensus are possible, but 'should be seen as temporary" (Mouffe 1999: 755). The task is to enhance 'passion' within politics and to realize that "agonistic confrontation is in fact [democracy's] very condition of existence" (Mouffe 1999: 756). To see democracy as agonism means to go beyond the *friend–enemy* thinking, and seeing the participant one heavily disagrees with or does not understand, as an adversary 'one can learn something from' (Mouffe 2000). This does not require the negligence of interests and power-mechanisms, but the need to respect differences and disagreements radically. According to Pløger (2004) the art of 'strife' is essential, allowing for a respectful way of disagreement.

> Following Newman again (2011): In this model, democratic agonism always takes place within the unacknowledged framework of the state, and it is unable to conceive of politics outside this framework. By situating democratic agonistic struggles primarily within the state and its parliamentary institutions, Mouffe leaves the actual political space of the state unchallenged.

As a reaction to changing societal circumstances, Ulrich Beck (1992, 1994) and others (Beck et al. 1994) have identified reflexive modernisation, a social theory, which emphasizes the new challenges current societies are experiencing due to the pressures exerted on existing institutions. A 'second modernity' emerges as the known rules of the first modernity are 'in flux' (Beck et al. 2003). Amin (2004) considers the capacity to change to be the basis. He acknowledges the fact that monitoring (and learning by monitoring) is a matter of developing a strategic and reflexive rationality (Lombardi 1994; Sabel 1994). **Reflexive planning** tries to capture this new modernity in planning. For example, De Roo and Porter (2007: 233) investigate the ways in which actors become continuously engaged in an "actor consulting model" for planning. Elsewhere, reflexive monitoring has been understood as "a participatory process of describing, evaluating, and reflecting on ongoing activities, designed to strengthen both the quality and impact of a project, concurrently, by feeding back into the project an understanding of its proceedings" (Grin and Weterings 2005: 5). Reflexivity has a strong temporal dimension it not only aims to solve present planning problems, but also imagines alternative trajectories for future action. It is seen as a new tool for generating critical knowledge and dialogue that can synthesise the perspectives of multiple actors in common understanding, within existing structural constraints and builds a collective imagination of alternative future possibilities. Reflexivity in planning focuses on 'projectivity', creativity and change; always bearing in mind that the future is uncertain, and that ready answers are not easy to come by (Lissandrello and Grin 2011).

> Considerations: it acknowledges the changing times, but it bases itself in rationality and dominantly with the government in the lead. Therefore, the results depend largely on the actors involved and if not directed strongly has the risk of being 'direction-loose'. Taking uncertainty as the interminable continuity a randomly chosen bunch of actors might reflex themselves into infinity. This approach has the risk in it of building a very accurate description of the changes in society and reflecting on this developments with a planning discourse of 'continuously involving stakeholders', which, in itself is valid, but leaves the question if these, even if carefully selected, stakeholders are capable of formulating responses to wicked problems.

However, reflexive modernity offers the framework of thought within which emerging debates take place within contemporary planning, concerning the fluidity of relations and interactions in planning processes, and the ways in which these processes influence future developments (e.g. Healey 2009; Hillier 2007). In this context Gunder (2011) calls for: "a critique that deconstructs both the planning discourses deployed, habitualised or otherwise derived, and their phantasmic affect upon both planners and the public. This is a call to challenge all positions which seek the security provided by the planner 'who knows'. This is a call to challenge what is often, at best, a mono-rational practice of orthodox and repetitive universal." In his pledge for post-anarchistic, or autonomous planning Newman emphasises the power of self-organising groups and organisations, planning for their own environments outside the governmental, political arena and creating herewith a *disordered order* of spaces that are 'becoming' (Newman 2011). In a debate provoking paper Boelens advocates planning to come from 'outside inward', led by actors out of the normal governmental planning arena (Boelens 2010). Miraftab describes the informal, insurgent, planning taking place in slums in South-Africa (Miraftab 2009) and Davy (2008) promotes unsafe planning on order to establish planning without tightening and dictating regulations. Gunder (2011) pledges to step away from the widespread code of what is unconsciously accepted 'good planning', positioning the planner as the one 'who knows', meanwhile, creating, following Davy: a "non-innovative state of mono-rationality". An alternative, which is capable to include wicked problems, looms when the fundamental properties of western planning mono-rationality (Davy 2008) are left behind, being:

- *'Playing by the rules'*, which in the case of wicked problems no longer rule;
- *'Repeat habitual prior experiences'*, which in wicked problem country is useless, because every time the problem appears to be unique; and
- Creating a *'non-innovative status quo'*, which is contra-productive if the wicked problem is 'already changing again'.

According to Davy, mono-rationality must be replaced by an 'unsafe' planning practice of poly-rationality, where liquid, turbulent or even wild boundaries of both planning thought and spatial territory can occur – literally, to do 'it' without the safety of a condom! This is a planning practice that takes risks, accommodates difference and encourages the new and creative. This type of practice is able to deal with uncertainties and the wicked character of problems such as climate change.

6.4.2 A Review of 2 Years of Planning Journals

The next step in the research is to examine whether unsafe, autonomous and poly-rational theories, concepts and strategies are discussed in the planning community, and if, in relation with the former, wicked problems are addressed. In order to gain insight about to what extent this specific part exists within the spatial planning debate, two volumes of four spatial planning journals have been analysed. The, in total 275 articles, which have been published in 2010–2011 in the Journals of Planning Theory (43), Planning Theory and Practice (34), The Australian Planner (45) and European Planning Studies (153), being the leading theoretical and practice oriented academic planning journals originating from two different continents, have been analysed. The articles were judged on criteria informing whether in the articles theories, concepts and strategies are discussed that potentially can deal with wicked problems. The following criteria have been distinguished:

- Integration (vs. thematic, specific, single subject): a wicked problem cannot be dealt with from a single narrow thematic perspective, because a singular solution for a problem that is wicked enables the problem to evolve into new forms the moment the thematic solution is executed. An integrated approach, in which themes and land-use functions are mutually connected and in which an area is approached as a whole, can deal much easier with unique, new and suddenly changing problems. Does the article approach problems in an integrative way or is it focusing on a specific theme or subject?
- Dynamic (vs. static): a division can be made in the aim of planning to stabilise the future or to emphasise dynamic environments, which need to be planned for and/or even need to be created. When wicked problems are taken into account spatial planning needs to recognise the existence of dynamic, continuous changing spatial settings and configurations. Does the article assume that planning

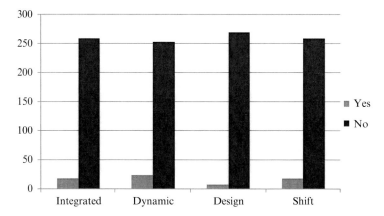

Fig. 6.2 Number of articles in Planning Theory, Planning Theory & Practice, Australian Planning and European Planning Studies (2010–2011) that do and do not reflect integration, dynamics, design and a paradigm shift

tries to continue the current state or focuses it on dealing with changing environments or subjects?
- Intervention (vs. regulatory): planning can be orientated on arranging general and objective regulations that prohibit or allow certain land-use, or it may aim for a deliberate change. A planning intervention can be realised through design. In general, if problems are wicked they normally are not dealt with by putting regulations in place, as these problems are essentially unique. Does the article discuss a design approach or a planning intervention or does it focus on describing regulations and institutions?
- Paradigm shift (vs. status quo): when problems are new, especially if they are wicked, a new planning paradigm may emerge. The identification of these types of problems at an early stage illuminates their existence in the first place. If so, the early stages of a paradigm shift are announced, even if there are only small rudiments of it visible. In most of the cases however, planning in its current state, a status quo, is described, which is less suitable in dealing with changing circumstances and wicked problems. Does the article describe planning as it is currently and/or was in the past or does the article focuses on identifying a paradigm shift.

Having analysed the 275 articles, addressing the question if they contain integrated or thematic, dynamic or stable, interventionist or regulatory and shift or status quo issues, the results are striking and shocking at the same time (Fig. 6.2). The fundamental properties of western planning mono-rationality are still around. Even stronger, articles that address dynamic, integrated, intervention topics and paradigm shifts are hardly found.

The conclusion may be drawn that a very small portion of current planning discourse acknowledges fundamental changes in society, the changes in the environment and the need to plan for wicked problems. However, the current debate is predominantly in the process of raising awareness and describing what is going on. It addresses the necessity to replace old rules for new ones, which can respond to more complex issues and are based on networks, interrelations and connections. There are only a few scholars (e.g. Newman's post-anarchism and Davy's unsafe planning), who discuss the necessity to start planning in a more 'non-linear' way. In this article the search for a planning theory dealing with wicked problems draws upon these scholars and will search where wicked problems are closest related to: complexity theory.

6.5 Exploring Complexity

In order to plan for wicked problems, and more specifically for climate adaptation, we need to take into account that it is likely that climate change will force (step) changes (Jones 2010), that climate change has locality specific characteristics and it requires to bridge impacts occurring over a wide time-range. Therefore, it is useful to explore the potential of complexity theory in three ways. Firstly, we need to understand complex (adaptive) systems, their non-linearity and the idea that small changes might have big impacts, as well as the existence of bifurcation points and tipping points. Secondly, we need to understand cities self-organising systems. And thirdly, we need to build upon the former to make this knowledge available for planning.

6.5.1 Complexity Theory

Many scholars studied the complexity and self-organisation of non-linear dynamic (or adaptive) systems. Amongst these are the works of Prigogine and Stengers (1984), Gleick (1987), Lewin (1993), Mitchell Waldrop (1992), Cohen and Stewart (1994), Kauffman (1995), which are further elaborated and explained by authors such as Johnson (2001), Miller and Page (2007), Johnson (2007) and Northrop (2011). Key concepts from complexity theory, which are seen as useful in a planning context, are the *self-organisation* of complex systems, the surge for an actor to attractors, depicting a *fitness landscape*, the change and transformation of a complex system in times of crisis and the existence of *bifurcation*, 'the point in time where for identical external conditions various possible structures can exist' (Allen 1996) and *tipping points*, 'the point at which the system 'flips' from one state to another' (Gladwell 2000).

Adaptation of (or within) the system is an internal process of *self-organisation*, which is the tendency in complex systems to evolve toward order instead of disorder (Kauffman 1993). The state of equilibrium is called attractor. Complex adaptive systems self-organise and adapt in order to remain within their current attractor. The system only shifts to other attractors (alternative states) after a shock that drives the system out of its current state (e.g. due to significant (or 'step') changes in climate). Major adjustments are needed and after the shock the system will self-organise to achieve those.

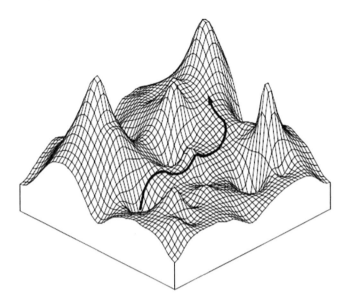

Fig. 6.3 Fitness landscape (Cohen and Stewart 1994) showing a complex system moving from a less favourable to a favourable position or attractor

The process this system goes through can be represented in the form of a *fitness landscape* (Fig. 6.3) (Mitchell Waldrop 1992; Langton et al. 1992). This fitness landscape includes favourable (the mountaintops) and less favourable (the valleys) positions. A complex system tends to move, while crossing less favourable valleys, to the highest possible position in the landscape, the attractor.

At the mountaintop, the adaptive capacity is highest, which allows the system to adapt more easily to changes in its environment. The pathway of the system is represented in Fig. 6.4. When a system self-organises it strives to reach a higher adaptive capacity by increasing order. When it reaches the mountaintop (B) it will continue to self-organise and increase order. However, by increasing order at this stage, adaptive capacity is decreasing, causing a less stable system (the state of fixed and unchangeable regulations and standards) and starts to move towards a new attractor. At this stage, the system is crossing the valley (from D to E) and searching for a new attractor, which can provide the system with renewed adaptive capacity. After reaching point E (a more chaotic state) two things can happen: the system dies (it didn't reach/find the other attractor) or it self-organises in a new way and starts to build up a transformed system by increasing its order again until it reaches its highest adaptive capacity (the mountaintop) again (B). Point E is defined as the bifurcation point, or: the point where the system fundamentally separates the pathway towards a new equilibrium from the one ending its existence ('die away'), also known as the tipping point, at which the system 'flips' from one state to another (Gladwell 2000).

These bifurcation, or tipping, points (1, 2) are the moments the system shifts from one state to another. In Fig. 6.5 these shifts are represented. At a certain point (1) the system in state x1 becomes less stable, for instance it is no longer capable of responding

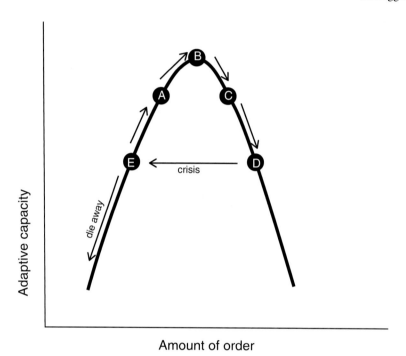

Fig. 6.4 Typical pathway of a complex system towards the mountain *top* (*B*), evolving towards instability (*D*), and dying or self-organising again (*E*) (After: Lietaer et al. 2009)

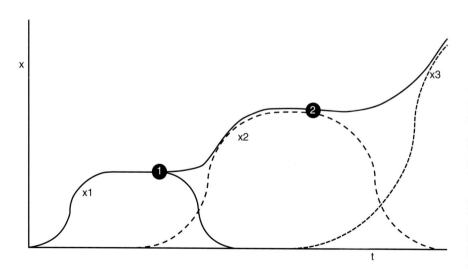

Fig. 6.5 Evolution of a complex system showing the rising and declining complexity of the system (*x*) in relation to time (*t*) and its bifurcation points (*1*, *2*) (After: Prigogine and Stengers 1984)

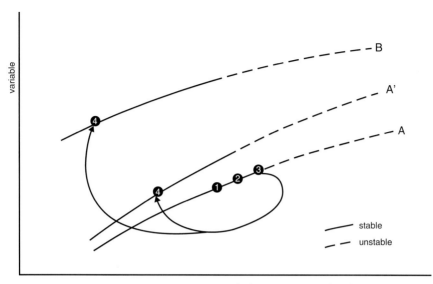

Fig. 6.6 Schematic behaviour of a complex system with a certain characteristic (*vertically*), developing over time (*horizontally*)

to the impacts of climate change. At this point the system (needs to) adapt quickly to the new environment and shifts into state x2, of higher complexity and order. If it fails to do so (the downward line) the system develops in a lower degree of complexity and cannot self-organise to deal with the new environment. It dies.

The behaviour of a complex system then consists of equilibrium phases and sudden changes (crises). In Fig. 6.6 (Timmermans et al. in print; after Prigogine and Stengers 1984) this process is visualised in the form of a slowly changing system A, which, due to external factors, such as climate change, can finally reach a less stable zone (2). Firstly, the system tries to maintain the equilibrium state A by suppressing change. At a certain moment, the system reaches a critical point (the bifurcation point) where it turns into instability (3). Here, at the edge of chaos, the system moves to a new equilibrium, A' or B (4).

•

6.5.2 Cities as Complex Systems

These theoretical concepts have been applied to cities. However, the majority of scholars (Allen 1996; Batty 2005; Portugali 2000, 2006, 2008) use complexity theory mainly to understand self-organising processes in cities through modelling of reality. Modelling remains a central activity at the intersection of complexity and spatial science (O'Sullivan 2004) but there is a growing concern about the implicit limitations of this 'orientation on modelling' as the relevance of the links between spatial and complexity theories becomes much wider (O'Sullivan et al. 2006). Still, the main attention in recent academic writings focuses on different kinds of computational representations of spatial analyses (O'Sullivan et al. 2006) and the representation in models through agent-based modelling or cellular automata (Crawford et al. 2005). The question is whether this *'mathematicalisation'* of the city offers more than only an understanding of self-organisation in cities, but merely supports cities in dealing with wicked problems, as it lacks the tools to influence the performance of the city. Spaces (and places) are, as described in Portugali (2006) mainly seen as an object to study, analyse, explain, understand, describe and model…. But this understanding is, to my knowledge, hardly used to inform planning and design processes on how to improve the quality of the city, or to better respond to and prepare for wicked problems.

6.5.3 Use of Complexity in Planning

As a bridge between the understanding of complexity in cities and planning for it, a key set of interrelated concepts that define a complex system (Manson 2001) can be helpful.

- At the core the *relationships* between its components and its environment, forming an ever-changing internal structure, determine the whole of the system. Due to the number and complexities of these relationships it is hardly possible to understand or predict the character of the whole system. Because of the wide array of complex internal relationships the system is in most cases able to respond to novel, external, relationships, but in the case there is no internal component capable of responding to novel external circumstances, which are for instance induced through climate change, this may end in a catastrophe for the system.
- The system exhibits *emergence*, e.g. the system wide characteristics stem from interactions amongst components (Lansing and Kremer 1993) and are thus much more than a simple addition of components qualities. It is difficult to anticipate change beyond the short term, because other components of the system adjust to the intervention in addition to other changes in the environment (Youssefmir and Huberman 1997). Any single change can have far-reading large-scale effects due to not understanding emergence from complexity (Lansing and Kremer 1993).
- A complex system performs *change and evolution* through three different capabilities: (1) self-organisation, e.g. the capability to adjust its internal structures to better interact with a changing environment; (2) Development of a dissipative structure, allowing the system to suddenly cross to a more organised state after being a certain period in a highly unorganised state (Schieve and Allen 1982); and (3) self-organised criticality, allowing the system to keep the balance between nearly collapsing and not doing so, caused by an internal restructuring, almost too rapid to accommodate, but necessary for survival (Scheinkman and Woodford 1994).

- Finally, *path dependency* defines the development of a system as 'a trajectory as function of past states' (O'Sullivan 2004). This may be true for most systems, Portugali demonstrated that in regards to planning the fact a plan has been released causes a *reverse* form of path dependency in the sense that the trajectory is a function of (not yet realised) future states (Portugali 2008). All of the former properties are found through the study of ecological and, to a lesser extent, economic systems.

The question, however, is whether we can use the knowledge derived from, mainly, ecosystems for artificial systems, such as cities. As demonstrated by Simon (1999, described in Portugali 2006) we can use the findings of natural science to apply in artificial systems, but only to a limited extent. As Portugali demonstrated (2000, 2008) social systems, such as cities and landscapes exhibit a dual complexity: the city as a whole is a complex adaptive system as is each of its parts (Portugali defines them as agents; e.g. human or organisational entities, Portugali 2008). This means that the whole can no longer be explained by the singular behaviour of individual components.

Learning from nature again, most systems performing *swarm behaviour* represent high resiliency, lessening the impact of uncertainties, complexity and change through the development of emerging patterns and structures (Van Ginneken 2009). Swarms (see Fisher 2009; Miller 2010) are self-organising systems in preparing and responding to changing circumstances, which is, according to Van Ginneken (2009), achieved through (1) the interactions taking place between a large number of similar and free moving 'agents', which (2) react autonomous and quick towards one another and their surrounding, resulting in (3) the development of a collective new entity and a coherent larger unity of higher order.

Swarm behaviour can be encouraged through increasing the adjustability of the building by programmatic labelling and tagging of building elements, enabling buildings to customise temporary desires or changing demands (Oosterhuis 2006, 2011),

The above forms the basis for developing the theory of Swarm Planning.

6.5.4 Proposition: Swarm Planning

The objective of this paper is to present the first contours and basic elements of Swarm Planning, which ultimately aims to increase the potential of a landscape or city to deal with wicked problems, such as climate change. Elaborating the above, this means that if the landscape could perform swarm behaviour, it increases its capacity to deal with uncertainty, complexity and change, hence dealing with wicked problems. Therefore, a planning theory that enables swarm behaviour to occur, supports landscapes to reach higher levels of adaptive capacity. This planning theory, Swarm Planning, needs to take at its core the dual complexity of the landscape and therefore to combine complex behaviour of the elements of the system and the complex adaptive behaviour of the system as a whole. And thus, Swarm Planning needs to actively intervene on both levels of the *'dual complex'* landscape.

6.5.4.1 The Intervention

At the level of the whole, an intervention needs to take place in order to start the swarm to behave in the first place. In current theory, tipping points are identified after they have occurred (Gladwell 2000) or identify the patterns that announce these points (Scheffer 2009), but they are not planned. In essence, it describes the process of an evolving system, becoming unstable, ends up in a crisis, 'tips' and transforms through self-organisation to another stable state. However, in the case of climate change, this system change preferably anticipates the actual change. Hence, an early intervention must allow the system to be able to 'flip'. We need to actively intervene in the system to start self-organising processes to anticipate the wicked problem. Hence, this requires an intervention point to get things started.

Obviously, the difficulty is to identify the location, the type and the actor to intervene. As demonstrated elsewhere (Roggema et al. 2012b) network theory holds the key to identifying the location. The type of intervention cannot be otherwise determined then through the local context (existing landscape combined with specific wicked problem). The actor identifies the point where and the type of intervention. The person or institution most eligible to decide upon this is the problem owner, not necessarily the government.

6.5.4.2 The Freedom to Emerge

The second level is the level of the parts (the elements in the landscape). At this level the components of the system need to make use of their joint capabilities to perform as a system as a whole. Only then, the system is able to produce swarm behaviour and achieve a higher adaptive capacity. Therefore, interacting relationships need to be provided with the qualities allowing them to develop emergent properties, to self-organise and to change (Manson 2001). The hypothesis is that if the landscape elements are attributed with the capabilities as described before, they will support swarm behaviour of the whole system. Once the individual components are attributed with these capabilities and free self-organisation will take place, the system will strive for the most optimal stable state (in general: the mountain top in the fitness landscape), which represents the highest adaptive capacity.

This theoretical proposition requires further research on the question how individual landscape elements can be attributed with qualities to allow them to perform emergent behaviour, self-organise and change. The first attempts to answer this question have been undertaken in the work of Kas Oosterhuis (2006, 2011), who attributed swarm characteristics to building elements, and by linking complex adaptive systems properties to landscape entities (Roggema 2011). However, further research is required in this area.

The proposition of Swarm Planning combines directive steering, in the form of an active design intervention (system level), with the freedom of individual landscape elements to shape (and self-organise) the system. The outcome of this process is fundamentally unpredictable, but this does not mean that we cannot be confident that the system, when performing swarm behaviour, reaches a higher adaptive capacity (or in complexity theory: reaches the top of the fitness landscape).

Part of a planning theory must be, in my opinion, besides a theoretical basis as presented above, a practical strategy and practical applications.

The theoretical basis has been used and translated into a practical approach (Roggema et al. 2012a) with the five layer strategy as the centrepiece, in which the first two layers identify the point of intervention, layer three arranges and defines the freedom to emerge and layers four and five allow for the individual components to self-organise.

The first practical applications of this theory have also been identified. In the work of Massoud Amin the principle of self-organisation in order to reach higher levels of agility in the energy network (Massoud Amin 2008a, b, 2009; Massoud Amin and Horowitz 2007) can be explained as an early form of Swarm Planning 'avant la lettre'. A second body of knowledge has been developed, designing 'swarm' landscapes for regional climate adaptation (Roggema 2008a, b; Roggema and Van den Dobbelsteen 2008).

6.6 Conclusion

In this chapter it has been demonstrated that current planning discourses are strongly focused on the government as major actor and rely mainly on existing rules, regulations and established procedures. Moreover, the academic debate, as represented in four international planning journals illuminates the scarcity of articles focusing on strategies and practices, which focus on interventions, incorporate a changing and dynamic future and emphasise a paradigm shift. This illuminates a gap in planning practice and theory. This gap will lead to suboptimal preparation for climate change impacts, both in adaptation and in energy supply. This may be seen as very risk-full, because, even if decision-makers should decide that climate change impacts need to become part of planning practice immediately, it will take a shift in the planning frameworks, which in itself takes amounts of time. This time lag might imply that is becomes too late to implement the required changes on time.

Complexity theory might open the opportunity to integrate the characteristics of wicked problems in spatial planning and as demonstrated in this article has been subject of debate to link complexity and city and geography, but unfortunately complexity theory is mainly used in a mathematical, modelling way to better understand self-organising processes in cities and not to identify design interventions or plans to increase the capability of cities (and landscapes) to prepare for the impacts of wicked problems (e.g. climate change). This leads to suboptimal preparation of communities in those cities and landscapes.

Therefore, in this article a proposition is launched to develop a planning approach, which can integrate complexity theory and uses it for planning and design purposes. Named 'Swarm Planning' is an attempt to do so and, learning from the fact that cities have been attributed with a dual complexity (Portugali 2000), it identifies two major levels of intervention: the whole system, a level at which a strategic intervention is required, and the level of the individual components, to which the properties of complex adaptive systems need to be attributed in order to allow free emergence. Both levels in conjunction are able to perform swarm behaviour, which improves resiliency through lessening the impact of uncertainties, complexity and change.

Compared with the way planning is practiced in many institutions, thinking in points enforcing change and free emergence, is the opposite of current practice. Generally, tipping points are not sought, but comprehensive developments are seen as the interventions and these comprehensive interventions are planned in great detail and for entire areas, not allowing them to develop freely. The aversion against tipping points, and surprises, and the willing to paternalise the entire planning process, including its detailed execution, is grounded in the political culture in many countries where risk has to be avoided and uncertainties or 'uncontrollabilities' must be abandoned. However, pursuing the existing (and historical) path-dependent political pathways will lead to 'more of the same' policy, which, and this is for certain, will not produce the planning interventions that are required to deal with the wicked problem of climate change.

And it is true, the results of Swarm Planning are, partly, unpredictable and this is, especially to the responsible decision-makers a danger, but it is also a *conditio sine qua non*. Because in Swarm Planning the new state of the system is undefined, and not possible to define either, there always is the danger of ending up with the wrong outcome, but continuing on the same pathway of not adjusting will end in repetition of history and this will certainly not bring the answers to fundamental different problems of the future. Having said this, there is a lack of understanding of what the future system, planned through Swarm Planning, may be, and more research can be carried out in this field. However, given the unpredictability future state of complex adaptive systems it can be questioned whether more understanding will shine brighter lights on the actual future of the system.

This leaves alone the potential of Swarm Planning to be used in landscape (and city) design. As the example design demonstrates it is very well possible to design a landscape by making use of dynamic and complex principles. Moreover, it illuminates the potential for a community to slightly move towards an adapted and safe state and at the same time to pursue their own desires in realising a future safe and resilient living environment. As this is only the first, implicit, design, the approach deserves further testing and application.

References

Allen PM (1996) Cities and regions as self-organising systems, models of complexity. Taylor & Francis, London/New York

Allmendinger P (2001) Planning in postmodern times. Routledge, London

Allmendinger P (2002a) Towards a post-positivist typology of planning theory. Plan Theory 1(1):77–99

Allmendinger P (2002b) Planning theory. Palgrave, New York
Amin A (2004) An institutionalist perspective on regional economic development. In: Barnes TJ, Peck J, Sheppard E, Tickell A (eds) Reading in economic geography. Oxford, Blackwell
Banfield EC (1973) Ends and means in planning. In: Faludi A (ed) A reader in planning theory. Urban and regional planning series, vol 5. Pergamon Press, Oxford/New York/Toronto/Sydney/Paris/Frankfurt
Batty M (2005) Cities and complexity, understanding cities with cellular automata, agent-based models, and fractals. The MIT Press, Cambridge, MA/London
Beauregard R (1996) Between modernity and post-modernity: the ambiguous position of US planning. In: Campbell S, Fainstein S (eds) Readings in planning theory. Blackwell, Oxford
Beck U (1992) Risk society: towards a new modernity (Trans. Ritter M). Sage, London
Beck U (1994) The reinvention of politics. In: Beck U, Giddens A, Lash S (eds) Reflexive modernization: politics, tradition and aesthetics in the modern social order. Polity, Cambridge, pp 1–55
Beck U, Giddens A, Lash S (1994) Reflexive modernization: politics, tradition and aesthetics in the modern social order. Polity, Cambridge
Beck U, Bonss W, Lau C (2003) The theory of reflexive modernization: problematic, hypotheses and research programme. Theory Cult Soc 20:1–33
Belin H (2008) The Rifkin vision. We are in the twilight of a great energy era. Eur Energy Rev, special edition, June 2008
Boelens L (2010) Theorizing practice and practising theory: outlines for an actor- relational – approach in planning. Plan Theory 9(1):28–62
Campbell CJ (1999) The imminent peak of world oil production. Presentation to a House of Commons All-Party Committee, 7 July 1999. http://www.hubbertpeak.com/campbell/commons.htm. Accessed 10 May 2010
Campbell CJ (2002a) Peak oil: an outlook on crude oil depletion. www.mbendi.com/indy/oilg/p0070.htm#27. Accessed 10 May 2010
Campbell CJ (2002b) Forecasting global oil supply 2000–2050. M. King Hubbert Center for Petroleum Supply Studies, Newsletter # 2002/3
Campbell CJ, Laherrrère JH (1998) The end of cheap oil. Global production of conventional oil will begin to decline sooner than most people think, probably within 10 years. Scientific American, 1998–03
Cates C (1979) Beyond muddling: creativity. Public Adm Rev 39(6):527–532
Cohen J, Stewart I (1994) The collapse of chaos: discovering simplicity in a complex world. Penguin Group Ltd., London
Commonwealth of Australia (2007) Tackling wicked problems; a public policy perspective. Australian Government/Australian Public Service Commission, Canberra
Crawford TW, Messina JP, Manson SM, O'Sullivan D (2005) Guest editorial. Environ Plan B 32:792–798
Davy B (2008) Plan it without a condom! Plan Theory 7(3):301–317
De Roo G (2006) Understanding planning and complexity – a systems approach. AESOP-working group complexity and planning, IIIrd meeting, Cardiff. http://www.aesop-planning.com/Groups_webpages/COMPLEX/cardiff/DeRoo.pdf. Accessed 8 Nov 2011
De Roo G, Porter G (2007) Fuzzy planning: the role of actors in a fuzzy governance environment. Aldershot, Ashgate
Dos Santos FT, Partidário MR (2011) SPARK: Strategic Planning Approach for Resilience Keeping. Eur Plan Stud 19(8):1517–1536
Dror Y (1964) Muddling through-'Science' or Inertia? Public Adm Rev 24:154
Dror Y (1973) The planning process: a facet design. In: Faludi A (ed) A reader in planning theory. Urban and regional planning series, vol 5. Pergamon Press, Oxford/New York/Toronto/Sydney/Paris/Frankfurt (reprinted, original publication in: Int Rev Adm Sci 29(1) (1963):46–58)
Farmer J (1993) A poststructuralist analysis of the legal research process. Law Libr J 85:391–404

Fisher L (2009) The perfect swarm, the science of complexity in everyday life. Basic Books, New York
Forester J (1989) Planning in the face of power. University of California Press, Berkeley
Gladwell M (2000) The tipping point. Little, Brown and Company/Time Warner Book Group, New York
Gleick J (1987) Chaos, making a new science. Penguin Books Ltd., Harmondsworth
Grin J, Weterings R (2005) Reflexive monitoring of system innovative projects: strategic nature and relevant competences. Paper prepared for the sixth open meeting of the human dimensions of Global Environmental Change Research Community, University of Bonn, October 2005
Gunder M (2011) Fake it until you make it, and then…. Plan Theory 10(3):201–212
Gunder M, Hillier J (2009) Planning, in ten words or less. Ashgate, London
Habermas J (1987) The theory of communicative action, vol 2. Polity Press, Cambridge
Habermas J (1993) Justification and application: remarks on discourse ethics. Polity Press, Cambridge
Healey P (1997) Collaborative planning: shaping places in fragmented societies. Palgrave, London
Healey P (2006) Collaborative planning: shaping places in fragmented societies, 2nd edn. Palgrave Macmillan, Basingstoke
Healey P (2009) In search of the "strategic" in spatial strategy making. Plan Theory Pract 10(4):439–457
Hillier J (2003) Agon'ising over consensus: why Habermasian ideals cannot be 'real'. Plan Theory 2(1):37–59
Hillier J (2007) Stretching beyond the horizon. A multiplanar theory of spatial planning and governance. Aldershot, Ashgate
Hillier J, Cao K (2011) Enabling Chinese strategic spatial planners to paint green dragons. Plan Theory 10(4):366–378
Innes J (2004) Consensus building: clarifications for the critics. Plan Theory 3(1):5–20
Innes J, Booher D (1999) Consensus building and complex adaptive systems – a framework for evaluating collaborative planning. APA J 65(4, Autumn):412–423
Innes J, Booher D (2004) Reframing public participation: strategies for the 21st century. Plan Theory 5(4):419–436
Jencks C (1987) Post-modernism, the new classicism in art and architecture. Academy Editions, London
Johnson S (2001) Emergence. The connected lives of ants, brains, cities and software. Scribner, New York/London/Toronto/Sydney
Johnson N (2007) Simply complexity, a clear guide to complexity theory. Oneworld Publications, Oxford
Jones R (2010) A risk management approach to climate change adaptation. In: Nottage RAC, Wratt DS, Bornman JF, Jones K (eds) Climate change adaptation in New Zealand: future scenarios and some sectoral perspectives. New Zealand Climate Change Centre, Wellington, pp 10–25
Kauffman SA (1993) The origin of order: self-organisation and selection in evolution. Oxford University Press, New York
Kauffman S (1995) At home in the universe. The search for laws of self-organisation and complexity. Oxford University Press, New York/Oxford
Langton CG, Taylor C, Farmer JD, Rasmussen S (1992) Artificial life II. In: Studies in the sciences of complexity. Proceedings vol 10. Santa Fe Institute, Redwood City
Lansing JS, Kremer JN (1993) Emergent properties of Balinese water temple networks: co-adaptation on a rugged fitness landscape. Am Anthropol 95(1):97–114
Lazarus R (2009) Super wicked problems and climate change: restraining the present to liberate the future. Cornell Law Rev 94:1053–1233
Lewin R (1993) Complexity, life at the edge of chaos. JM Dent Ltd., London
Lietaer B, Ulanowicz R,Goerner S (2009) Options for managing a systemic bank crisis, S.A.P.I.EN.S [Online], 2.1 |, Online since 06 April 2009, Connection on 03 October 2012. URL: http://sapiens.revues.org/747

Lindblom CE (1959) The science of "Muddling Through". Public Adm Rev 19(2):79–88
Lissandrello E, Grin J (2011) Reflexive planning as design and work: lessons from the port of Amsterdam. Plan Theory Pract 12(2):223–248
Lombardi M (1994) L'evoluzione del distretto industriale come sistema informativo: alcuni spunti di riflessione. L'Industria 15(3):523–535
Manson SM (2001) Simplifying complexity: a review of complexity theory. Geoforum 32:405–414
Massoud Amin S (2008a) Resilience and self-healing challenges: present-possible futures. In: CRITIS'08, Third international workshop on critical information security, Frascati-Rome. http://critis08.dia.uniroma3.it/pdf/CRITIS_08_9.pdf. Accessed: 8 Nov 2011
Massoud Amin S (2008b) The smart self-healing electric power grid: challenges in security and resilience of energy infrastructure. In: Proceedings of the 2008 IEEE Power Engineering Society general meeting, Pittsburgh, July 2008. http://panda.ece.utk.edu/w/images/e/e7/Amin_IEEE_PES_GM_July_2008.pdf. Accessed 8 Nov 2011
Massoud Amin S (2009) Smart grid: opportunities and challenges – toward a stronger and smarter grid. Keynote at the Smart grid workshop sponsored by Sandia National Laboratories at the MIT energy conference, Cambridge, MA, 6 March 2009. http://panda.ece.utk.edu/w/images/9/92/2009_SNL_Amin_MIT_Energy.pdf. Accessed 8 Nov 2011
Massoud Amin S, Horowitz B (2007) Toward agile and resilient large-scale systems: adaptive robust national/international infrastructures. In: Flexibility with business excellence in the knowledge economy, pp 247–265, November 2007, ISBN: 81-903397-7-X. http://panda.ece.utk.edu/w/images/1/1a/Amin_Agile_Resilient_Systems.pdf. Accessed 8 Nov 2011
McLoughlin B (1969) Urban and regional planning: a systems approach. Faber & Faber, London
Miller P (2010) The smart swarm. The Penguin Group, New York
Miller JH, Page SE (2007) Complex adaptive systems. An introduction to computational models of social life. Princeton University Press, Princeton/Oxford
Miraftab F (2009) Insurgent planning: situating radical planning in the Global South. Plan Theory 8(1):32–50
Mitchell Waldrop M (1992) Complexity. The emerging science at the edge of order and chaos. Simon and Schuster Paperbacks, New York/London/Toronto/Sydney
Mouffe C (1993 [2005 edition]) The return of the political. Verso, London
Mouffe C (1999) Deliberative democracy or agonistic pluralism. Soc Res 66(3):745–758
Mouffe C (2000) The democratic paradox. Verso, London
Mouffe C (2005) On the political. Routledge, New York
Newman S (2011) Post-anarchism and space: revolutionary fantasies and autonomous zones. Plan Theory 10(4):344–365
Northrop RB (2011) Introduction to complexity and complex systems. CRC Press/Taylor & Francis Group, Boca Raton/London/New York
O'Sullivan D (2004) Complexity science and human geography. Trans Inst Br Geogr 29:282–295
O'Sullivan D, Manson SM, Messina JP, Crawford TW (2006) Guest editorial. Environ Plan A 38:611–617
Oosterhuis K (2006) Swarm architecture II. In: Oosterhuis K, Feireiss L (eds) Game, set and match ii, on computer games, advanced geometries and digital technologies. Episode Publishers, Rotterdam
Oosterhuis K (2011) Towards a new kind of building, a designer's guide to nonstandard architecture. NAi Uitgevers, Rotterdam
PBL, KNMI, WUR (2009) News in climate science and exploring boundaries, a policy brief on developments since the IPCC AR4 report in 2007. PBL publication number 500114013, Bilthoven
Pløger J (2004) Strife: urban planning and agonism. Plan Theory 3(1):71–92
Portugali J (2000) Self-organisation and the city. Springer, Berlin/Heidelberg/New York
Portugali J (2006) Complexity theory as a link between space and place. Environ Plan A 38(4):647–664
Portugali J (2008) Learning from paradoxes about prediction and planning in self-organising cities. Plan Theory 7(3):248–262

Prigogine Y, Stengers I (1984) Order out of chaos. Man's new dialogue with nature. Bantam Books, Inc., New York

Ramirez R, Selsky JW, van der Heijden K (eds) (2008) Business planning for turbulent times, new methods for applying scenarios. Earthscan, London/Sterling

Richardson K, Steffen W, Schellnhuber HJ, Alcamo J, Barker T, Kammen DM, Leemans R, Liverman D, Munasinghe M, Osman-Elasha B, Stern N, Wæver O (2009) Climate change – synthesis report: global risks, challenges and decisions, Copenhagen 2009. University of Copenhagen, Copenhagen

Rifkin J (2002) The hydrogen economy: the creation of the world-wide energy web and the redistribution of power on earth. Penguin Group (USA) Inc., New York

Rittel H, Webber M (1973) Dilemmas in a general theory of planning. Policy sciences, vol 4, Elsevier Scientific Publishing Company, Inc., Amsterdam, pp 155–169, 1973 (reprinted in Cross N (ed) Developments in design methodology. Wiley, Chichester, pp 135–144, 1984)

Roggema R (2008a) The use of spatial planning to increase the resilience for future turbulence in the spatial system of the Groningen region to deal with climate change. In: Proceedings UKSS -conference, Oxford

Roggema R (2008b) Swarm Planning: a new design paradigm dealing with long term problems associated with turbulence. In: Ramirez R, Selsky JW, Van der Heijden K (eds) Business planning for turbulent times, new methods for applying scenarios. Earthscan, London/Sterling, pp 103–129

Roggema R (2011) Swarming landscapes, new pathways for resilient cities. In: Proceedings 4th international urban design conference 'Resilience in Urban Design', Surfers Paradise

Roggema R, van den Dobbelsteen A (2008) Swarm Planning: development of a new planning paradigm, which improves the capacity of regional spatial systems to adapt to climate change. In: Proceedings world sustainable building conference (sb08), Melbourne

Roggema R, Van den Dobbelsteen A, Kabat P (2012a) Towards a spatial planning framework for climate adaptation. SASBE 1(1):29–58

Roggema R, Vermeend T, van den Dobbelsteen A (2012b) Incremental change, transition or transformation? Optimising change pathways for climate adaptation in spatial planning. Sustainability 4(10):2525–2549

Sabel CF (1994) Learning by monitoring: the institutions of economic development. In: Smelser N, Swedberg R (eds) Handbook of economic sociology. Princeton University Press, Princeton

Scheffer M (2009) Critical transitions in nature and society. Princeton University Press, Princeton/Oxford

Scheinkman JA, Woodford M (1994) Self-organised criticality and economic fluctuations. Am Econ Rev 84(2):417–421

Schieve WC, Allen PM (eds) (1982) Self-organisation and dissipative structures: applications in the physical and social sciences. University of Texas Press, Austin

Schwank T (1965) Sensitive chaos, the creation of flowing forms in water and air. Rudolf Steiner Press, Forest Row

Sergeev V, Roggema R, Artrushkin V, Mallon W, Alekseenkova E (2009) What is the price to use gasoline in contemporary society? The Groningen experience. In: Roggema R (ed) INCREASE 2, INternational Conference on Renewable Energy Approaches for the Spatial Environment, conference proceedings. Province of Groningen, Groningen

Simon HA (1999) The sciences of the artificial. MIT Press, Cambridge

Sommerkorn M, Hassol SJ (eds) (2009) Arctic climate feedbacks: global implications. WWF International Arctic Programme, Oslo

Timmermans W, van Dijk T, van der Jagt P, Onega Lopez F, Crescente R (in print) The unexpected course of institutional innovation processes. Inquiry into innovation processes in land development practices across Europe. Int J Des Nat Ecodyn

Tin T (2008) Climate change: faster, stronger, sooner, an overview of the climate science published since the UN IPCC fourth assessment report. WWF European Policy Office, Brussels

Van Dam F, Noorman K-J (eds) (2005) Grounds for change: bridging energy planning and spatial design strategies. Charrette report. Grounds for Change/IGU, Groningen

Van Ginneken J (2009) De kracht van de zwerm. Uitgeverij Business Contact, Amsterdam/Antwerpen

VROM-raad (2007) De hype voorbij, klimaatverandering als structureel ruimtelijk vraagstuk. Advies 060. VROM-raad, Den Haag

Youssefmir M, Huberman BA (1997) Clustered volatility in multiagent dynamics. J Econ Behav Organ 31(1):101–118

Chapter 7
Swarm Planning Methodology

Rob Roggema

Contents

7.1	Introduction	142
7.2	The Whole and the Parts	142
7.3	Swarm Planning Framework	143
	7.3.1 The Layer Approach	143
	7.3.2 Use in Practice	145
	7.3.3 Application in Groningen Province	146
7.4	Design Charrettes	150
	7.4.1 Involvement Through Design	152
	7.4.2 The Groningen Charrettes	155
	7.4.3 The Victorian Design Charrettes	157
	7.4.4 Key Success Factors	159
7.5	Swarm Planning Experiment	160
7.6	Conclusion	164
References		164
Websites		166

Abstract In this chapter the question how to develop a spatial plan that is able to deal with the unpredictable impacts of climate change is explored. Based on the layer-approach a Spatial Planning Framework for Climate Adaptation is developed, consisting of five layers, each with their specific time-rhythm. All spatial elements can be connected to one of the layers, depending on the pace of change they tend to change. Subsequently the five layers can be used in practice to create a climate proof spatial plan. The process in which the development of a climate proof plan can be best developed needs to appeal creativity and future thinking. Two processes

---

R. Roggema (✉)
The Swinburne Institute for Social Research, Swinburne University of Technology,
PO Box 218, Hawthorn, VIC 3122, Australia
e-mail: rob@cittaideale.eu

are extremely suitable for developing these kinds of plans: Design Charrettes and the COCD-method. The success of design charrettes lies in the successful use of local expertise and the collective creativity to visualise on maps the desired climate proof future. The COCD-method is successfully used in the Swarm Planning Experiment, creating specific Swarm Plans for the Eemsdelta region in the Netherlands.

Keywords Swarm planning • Methodology • Layer-approach • Design-charrette • COCD-box

7.1 Introduction

The theory as described in Chap. 6 offers the contours for a methodological approach how to create swarming plans. In this Chap. 2 methodological aspects are highlighted: the content, e.g. what is the method that create high quality swarm plans, and the process, e.g. which working methods and planning processes can be ideally used within which swarm plans can be created? The chapter is divided in these two parts. The content part starts with a brief description of the city of two complexities (Sect. 7.2) and follows up with the development of the Swarm Planning Framework (Sect. 7.3). The process part highlights the benefits of the Design Charrettes (Sect. 7.4) and ends with a description of the SASBE special session, in which the Swarm Experiment took place (Sect. 7.5).

7.2 The Whole and the Parts

Key part of the Swarm Planning Theory is that complexity insights are used to create plans that meet the characteristics of cities and landscapes. Portugali (2000) found that the urban system as a complex system consists of two complexities. The city as a whole functions as a complex adaptive system and can therefore accordingly be approached in the form of directive steering through an active design intervention (system level). The second complexity is found at the level of the individual spatial elements, which each perform as a complex adaptive system, too. This allows these individual landscape elements the freedom to together self-organise and shape the system. The results in terms of how a future landscape looks like when directed by intervention at the system level in combination with the freedom of individual elements to self-organise, is fundamentally unpredictable. However, Chap. 6 has outlined that the system, when performing this kind of 'swarm' behaviour, reaches a higher adaptive capacity (Roggema 2012).

This theoretical basis has been used and translated into a practical approach with the five layer strategy as the centrepiece (Roggema et al. 2012), in which the first two layers identify the point of intervention, layer three arranges and defines the freedom to emerge (hence, layer one, two and three plan mainly for the system level) and layers four and five allow for the individual components to self-organise.

7.3 Swarm Planning Framework

The development of a spatial planning framework for climate adaptation (Roggema et al. 2012) finds its foundation linking the time dynamics of different elements of complex adaptive systems with different spatial 'layers', as identified in the 'layer approach'.

7.3.1 The Layer Approach

The layer approach (Frieling et al. 1998) defines three layers for different timeframes or 'rhythms'. The rhythm of the first layer (water and soil, the underground) is centuries. To a large extent the water system and the soil determine possible uses of land, including the spatial elements that can or cannot function in a certain area. The second layer (networks) has a rhythm of approximately 100 years. Transport and energy networks yet also ecology belongs to this layer, often represented as linear elements. The third layer (occupation) is linked with a timeframe of 20–50 years (one generation). The patterns derived from human use of the landscape are culturally determined: heritage, agriculture, economic functions, recreation and living. According to De Hoog et al. (1998) a fourth layer, 'the public domain', can be added to the original three. This fourth layer is meant to provide impulses at strategic points (nodes, centres) in the urban system (e.g. focal points) and is considered to have a time rhythm of 5–20 years.

The layer approach is extremely helpful in integrating long-term changes, such as climate change, because it enables the connection of different timehorizons. Each layer defines a different time rhythm, hence it can be used to allocate spatial elements according their specific timeframe over with they tend to change. The three layers of Frieling et al. (1998) with the addition of De Hoog et al.'s fourth layer (1998) have been added with a fifth layer (Roggema et al. 2012). This layer ('unplanned space'), which has the shortest time rhythm (1–5 year), aims to include highly dynamic, emergent properties of systems, we propose a new, fifth layer. The layer is process oriented, as it illustrates starting-points of developments (emergent places) and the surrounding unplanned space.

The five layers (Fig. 7.1) form the basis of the planning framework for climate adaptation and are capable to cover time dynamics of every spatial element.

The dynamics, the time rhythm and the changeability of the layers have been defined as follows (Roggema et al. 2011a):

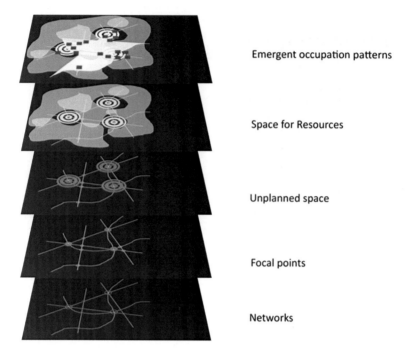

Fig. 7.1 Adjusted layer theory for climate adaptation planning (Roggema et al. 2011a, b)

- **Layer One**: *Networks* are adjustable, but remain steady over longer periods. To build a new network (a road, electricity grid) takes up to 10 years. Once these networks have been built, they hardly change a 100 years after. The ecological network is specific, as it can be manmade but generally emerges naturally. These networks remain for a long period. The transport-, water-, ecology- and energy-network are part of this layer;
- **Layer Two**: Network linkages determine the *focal points* and they can change, but also stay the same over longer periods. Changes in network patterns, which may occur every 5–20 years, direct these points. The most significant nodes, e.g. where different and intense networks cross one another, belong to this layer. These nodes are the public spaces and landmarks in the system, where interactions take place and developments can emerge. These points can be seen as bifurcation points, the points and moments where a system transforms to another stable state (amongst others: Portugali 2000). Here, spatial interventions and impulses start processes and developments that are capable of anticipating future changes and that increase the adaptive capacity of a system. By actively directing the nodes in the networks, processes of action and reaction will start and individual actors in the system will 'automatically' start to adjust in the most optimal way.
- **Layer Three**: *Unplanned space* is highly dynamic, because change needs to be possible during a hazard when this space needs to change functionality and temporarily be available. The area surrounding the focal points (layer

two) remains free of any specific function but can be occupied when a sudden event happens. For instance, when heavy rainfall causes flooding, these unplanned spaces are the areas for inundation and temporary use. Unplanned space gives room to processes of self-organisation, in which feedbacks lead to new standards of a more flexible system. Despite the fact that it is common sense to approach spatial planning in this way (e.g. keeping spatial options open for unexpected future change), the impacts of climate change are often sudden, disasters become more severe and this requires larger spaces than are provided in regular planning processes (if this is being done in current practice at all). The unplanned space will, after having been 'used' to cater for sudden climate events, return to its unplanned status and will remain unplanned until the moment it is needed again. The area changes back and forth during 1–10 years.

- **Layer Four**: The underground determines locations for *natural resources*, such as food, water, energy and nature and preserves them on the longer term. Based on existing soil and water conditions, areas for the production of food, drinking water and energy as well as the location of nature reserves can be determined. These locations are long lasting, steady and will change only after rigorous changes in circumstances, e.g. long droughts, cold periods or heat. These types of changes only occur over centuries, if at all.
- **Layer Five**: In the fifth layer *occupation patterns emerge* over time and adjust to changing circumstances. This layer is characterised as 'slow pace dynamic'. The space required to deal with climate hazards (floods, fires, heat and droughts) provides safe living environments, different mixes of functions in the landscape and in the city. They offer specific identities, landmarks and entities. It will change if new demands self-organise into new patterns, but these patterns are usually upcoming or declining over periods of 3–10 years.

7.3.2 Use in Practice

When the framework is used in practice, the five layers will not be designed simultaneously. The proposed way to use the framework is in a sequential process of several iterations (Fig. 7.2), of which the first one is mainly analytical and aiming to identify the focal points. The other stages in the process then design unplanned space (iteration two), and subsequently spaces for resources and occupation. As shown in Fig. 7.6, this must be seen as a cyclical process: in the first iteration layer one and two are connected, while in the second iteration layer three is connected to both layers one and two. The process repeats itself in iteration three, where layer four is connected to layers one, two and three, and so forth. This cyclical, iterative process facilitates setting priorities, especially by choosing the most important focal points first and then designing the required unplanned space around it. The rest will follow as a result of these first choices. Now that this framework has been defined, it can be used to develop climate-adaptive spatial plans.

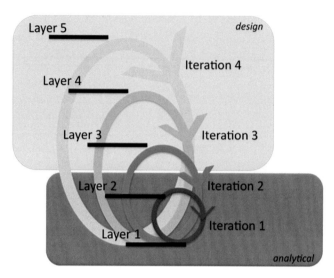

Fig. 7.2 The framework in practice

7.3.3 Application in Groningen Province

The spatial planning framework was used to develop a regional adaptive plan for the Dutch province of Groningen. The spatial planning framework discussed in Sect. 7.6 was used as a step-by-step approach to construct an alternative regional plan, aiming to improve the adaptive capacity of the area.

In the first layer the major networks were determined (Fig. 7.3). Bundles of networks, where roads, railways, energy networks, ecological corridors and waterways are combined, function as the main drivers of activities. Many of these networks are flexible and contain back-up structures, allowing the system to keep operating when parts of the network fail.

The first iteration illuminates the identification and planning of the crucial focal points. In the focal points where bundles of networks intersect (Fig. 7.4) interactions are likely to be more intense and auto-develop emergent processes. In these nodes people come together and exchange ideas. And here they anticipate and respond to future changes. When a system transformation is required to increase resilience, this is likely to start and happen here. Likewise, interventions consciously planned to enhance system change are likely to be most successful in these locations. The identification of these 'places of intervention' requires further elaboration, since they play a strategic role in the entire framework.

The focal points determine the places where emergent processes may start. However, these self-organising processes require unplanned space (Fig. 7.5) around them, allowing for free developments and occurrence of feedback mechanisms. These spaces are identified and designed in iteration 2. For instance, in case of flooding, these areas around focal points can transform into water storage basins. In case of a heat wave, these areas can be used to provide cooling shelters.

7 Swarm Planning Methodology

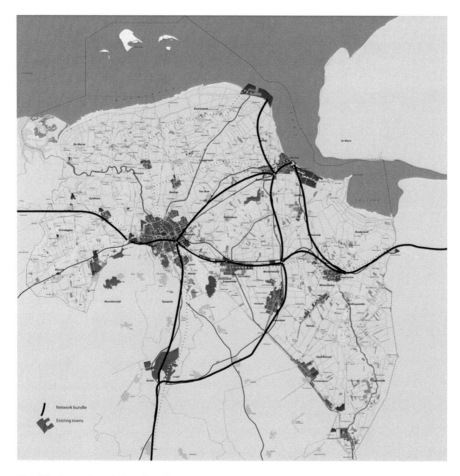

Fig. 7.3 Layer 1, main bundles of networks in the province of Groningen

Around a focal point the first zone of influence, i.e. where transformations take place most immediately is identified. Beyond this zone unplanned space is reserved to accommodate uncertain developments. If the distance between two focal points is long enough and there is space on or along the network bundle, new emergent places may develop. These emergent focal points will subsequently develop a zone of influence and unplanned space around them. Through short-term adaptation these zones connected to network bundles will be highly dynamic and capable of changing and dealing with unexpected changes.

In the areas outside the highly dynamic zones, the topography, soil and water system determine the most optimal locations for food and energy supply, water storage and ecological structures (Fig. 7.6) in the third iteration of the design process. The patterns occurring in these spatial reserves for natural resources are related to the spatial densities in the landscape: wide and open versus small and condensed. Much space is allocated for the storage of water, because both agriculture and

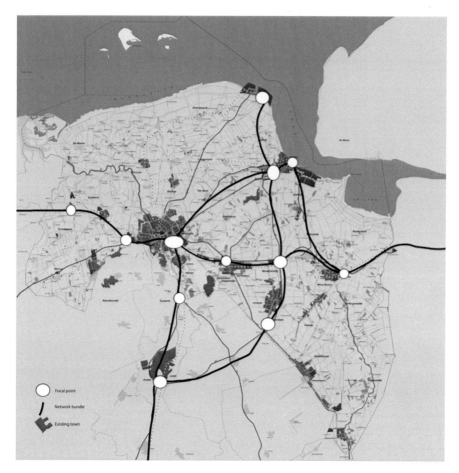

Fig. 7.4 Layer 2, focal points in the Province of Groningen

humans require a lot and in the future water will become scarce in summer. As a result of the allocation of spatial reserves the area will be less vulnerable and more robust to external shocks and unprecedented impacts of climate change. In rural areas the function mix ensures a great diversity and flexibility, allowing for easy adjustment and self-healing capacity in case the environment changes.

The final step in constructing a climate-adaptive regional plan, iteration four, incorporates the increase of functional differences in urban areas and the arrangement of safe areas to live (Fig. 7.7). In the Groningen case study 'safe living' mainly implies a thorough coastal defence. In the plan this was taken care of by introducing a defence zone, consisting of multiple dikes and an intermediate flood mitigation zone. Inland from this zone safety levels are much higher than current standards. Hence, the region becomes a more robust and less vulnerable system, which will also have self-healing capacity if one of the dikes breeches.

7 Swarm Planning Methodology

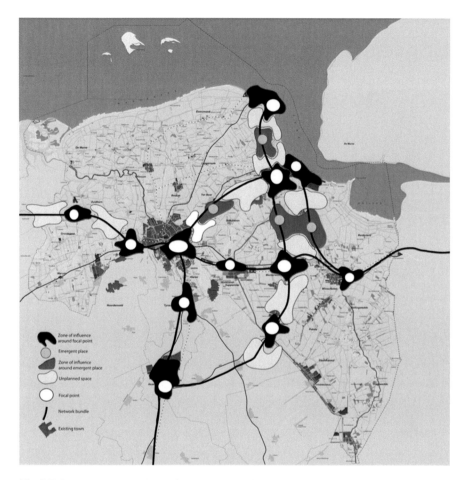

Fig. 7.5 Layer 3, unplanned space in the Province of Groningen

The other aspect of the fifth layer – emergent occupation patterns – will increase the diversity and flexibility of the system, allowing for coexistence and new standards to emerge. A mix of functions in intense urban areas will stimulate interaction within and between communities. This mix of people, social groups and urban functions increases the capability to adapt quickly and easily, enhancing the adaptive capacity.

The benefits of using the Swarm Planning Framework are threefold:

1. It enabling spatial systems to adapt to climate change;
2. It includes a range of time dimensions through structural use of the five designated layers;
3. It enriches the pallet of spatial interventions and elements that can be used to design a climate adaptive spatial plan.

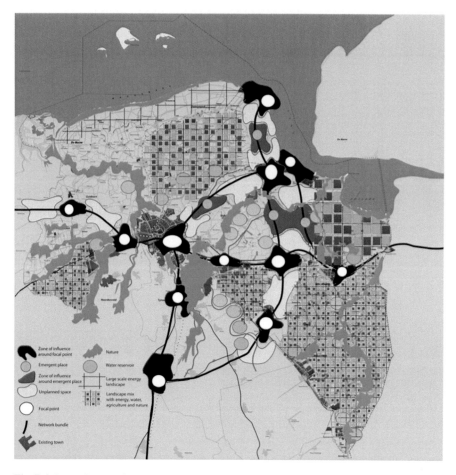

Fig. 7.6 Layer 4, space for natural resources in the Province of Groningen

7.4 Design Charrettes

Specific processes are required to be able to develop climate adaptive plans, or using the Swarm Planning Framework. In contrast, regular planning processes:

- Often copy former planning processes and therefore come up with former solutions, even if problems are new;
- Are dominated by paperwork and traditional meeting formats. In this atmosphere solutions and ideas are often less innovative;
- Tend to involve the 'usual suspects'; e.g. if a design charrette is organised designers are invited, if an agricultural expert meeting is organised farmers sit around the table. This often leads to repetitive outcomes or outcomes that could have been expected beforehand;

7 Swarm Planning Methodology

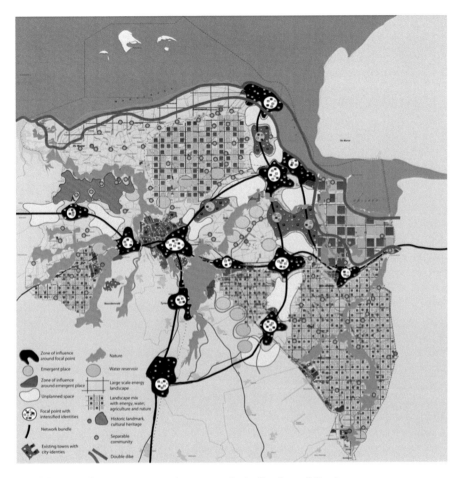

Fig. 7.7 Layer 5, emergent occupation patterns in the Province of Groningen

- Are conducted in a meeting room, leaving participants the chance to get distracted and are off-attention from the issue at stake;
- The end results are not celebrated, but seen as just another normal product 'we always come up with'. This does not motivate people during the process to make something special of it.

Three underlying problems occurring in regular planning processes prevent the development of innovative plans, dealing with the complex issue of climate adaptation:

1. Spatial planning and design aims to provide solutions for relatively straightforward, 'tame' problems (Conklin 2001), while climate adaptation is seen as a 'wicked' problem (Rittel and Webber 1973; VROM-raad 2007; Commonwealth of Australia 2007) for which no definitive solution exists because these problems are dynamic and ever changing (Roggema et al. 2012).

2. The second problem lies in the way stakeholders are involved in regular planning processes. In the majority of cases stakeholders are 'consulted', which means they are approached with an already well thought through and well-developed design proposal. The role left for the stakeholders is in general to accept or reject such proposals. Real influence or a contribution in the form of suggestions is often not possible, nor welcomed.
3. The third problem is that different stakeholders are separated in different processes. The stakeholders involved in climate adaptation (the 'environmentalists') differ from the ones involved in spatial planning (the 'designers'). Depending on the subject of the process, specific sub-groups of stakeholders show up *and* are invited. Exchange and learning rarely takes place. There is no joint 'framework of operations'.

These three problems inhibit the inclusion of the Swarm Planning Frameworks and realisation of climate adaptive plans. Existing practice separates climate adaptation and urban planning, prevents stakeholders from early involvement with the potential of dividing different stakeholder groups denying the opportunity of considered and well-accepted plans. Therefore, an alternative method is required, which can function as a platform for sharing the climate adaptive vision. Such a method has been found in the form of *design charrettes*.

Not only is climate adaptation seen as a wicked problem, design and planning problems are also identified as being wicked (De Jonge 2009). The combination of both wicked problems of design and climate adaptation is one of the reasons why integration of climate adaptation in spatial designs is proven to be difficult and why an alternative approach needs to be found. The charrette approach, which does not focus on the one final solution for the problem, offers the space within which *'an interactive exploration of potential strategies aiming to facilitate a future spatial development towards a status of improved adaptation to the impacts of climate change'* can take place. In this space, one-on-one technical-rational solutions are rare and different future thinking techniques are explored. As climate change predictions come with a broad margin of certainty, so do designs. A wide range of designs is able to provide improvements for one single problem.

7.4.1 Involvement Through Design

The charrette originates from France. At the end of the nineteenth century the Architectural Faculty of the *Ecole des Beaux-Arts* issued problems that were so difficult few students could successfully complete them in the time allowed. As the deadline approached, a pushcart (or 'charrette' in French) was pulled past students' workspaces in order to collect their final drawings for jury critiques while students frantically put finishing touches on their work. To miss the charrette meant an automatic grade of zero. The NCI defines the charrette as: "a collaborative design and planning workshop that occurs over 4–7 consecutive days, is held on-site and

7 Swarm Planning Methodology

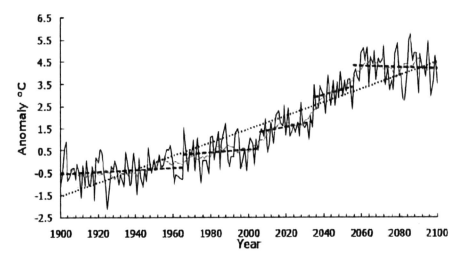

Fig. 7.8 Step and trend analysis, based on dummy data, illustrating one of many analyses showing the 'staircase' behaviour of climate change (Jones 2011)

includes all affected stakeholders at critical decision-making points" (Lennertz and Lutzenhiser 2006). Building on this Condon formulates it as: "a time-limited, multiparty design event organised to generate a collaborative produced plan for a sustainable community" (Condon 2008). As highlighted elsewhere a wide variety of design charrettes has been executed, especially in North Western Europe and North America (Roggema et al. 2011a, b).

Dealing with the wicked character of a changing climate, requires the generation of new knowledge. In dealing with unexpected step changes, which, as an example can be illustrated using temperature data of recent decennia (Fig. 7.8) (Jones 2011) existing knowledge does no longer satisfy as it produces the same solutions for fundamental new problems as it id for past problems. This generation of *new knowledge* helps to adapt to a future, not yet particularly clear, environment as it supports the design and the *transformation* of regions towards climate proof urban and regional areas. People need to become *engaged* to develop this knowledge and these future visions. In general, people have difficulties in dealing with wicked problems and uncertainty, even if they are professionals. Through direct involvement in the design process the capacity of people to deal with uncertainty of future climate change increases.

Hence, in order to develop a successful design charrette the following aspects are brought together:

1. *Knowledge creation*: Generation of new tacit knowledge. Tacit knowledge, opposite to explicit knowledge (transmittable in formal, systematic language), has a personal quality, which makes it hard to formalise and communicate. Tacit

knowledge is deeply rooted in action, commitment, and involvement in a specific context (Polanyi 1966, cited in Nonaka 1994). This type of knowledge creation, which is useful in dealing with far from logic, formal and systemic phenomena such as climate change, can be encouraged through (1) construction of the field: building self-organising team, (2) sharing experience, (3) conceptualise, (4) crystallise in the form of product (facilitated by encouraging experimentation) and (5) justification. In the process, in order to create successfully new knowledge, there needs to be a certain level of creative chaos and redundancy of information, making it possible to provide new information. A good example of both involvement and new knowledge creation has been carried out in the city of Tromsø, Norway, where the design is seen as a (creative) experimentation, involving people and defining planning as an inclusive process, whilst breaking with institutionalised practices (Nyseth et al. 2010);

2. *Governance*: A traditional 'top-down' policy approach, in which the State directs, manages and takes care of all citizens is, given the wicked character of climate change and the urge for enhancing adaptive capacity in society, suboptimal or even contra-productive. The roles of and relations between State and its citizens, more often organised in strong network relationships within and outside specific societal groups, needs to be redefined. Adaptive capacity can be enhanced if flows of resources and information between individual elements in the network as well as outside these networks can flow freely and a mutual relation between State and society is established (Adger 2003);

3. *Transformation*: It seems evident that with sudden and unexpected changes in global and regional climates, urban regions and landscapes need to undergo a transformation in order to be able to deal with those surprising circumstances. Such transformations are not new. Larger cities, in preparation for mega-events did transform, using marketing tools and urban planning and design as instruments to shape an image of the region after transformation. Good examples are Glasgow in its preparation to become Europe's Cultural Capital (García 2005) and Barcelona, preparing for the'92 Olympics [www.mt.unisi.ch]. Several basic drivers are identified to be able to enforce transformations of urban regions. Urban transformations, as outlined in Chap. 4 find their origin in (1) pressure from the outside landscape (the general context), (2) dissatisfaction with the current, stable regime, and (3) start as novelties and niche developments, which ultimately lead to breakthroughs in the existing regime (Geels 2002, 2005, 2011).

In conducting design charrettes Condon (2008) defines nine general rules for a good process. The four we acknowledge as the most significant are:

1. Design with everyone: Despite the fact that becoming a designer requires thorough training and very specific skills, the design process as undertaken during charrettes is integrative and contains a variety of possible solutions. This is partly an intuitive and judging activity, which makes it accessible for many individuals. In this sense, everyone is a designer;

2. Start with a blank sheet: If the group of participants are standing around the table, on which a large map of the site is laid down, the simple action to overlay this map with a blank piece of transparent paper will do. The invitation and the challenge lie before all. Everyone is invited to fill in the future and a shared vision will, in the hours to follow, fill up the formerly empty paper;
3. Provide just enough information: Too much information causes decision paralysis and too little produces bad proposals. Just enough is mainly arranged through the expertise of the participants and will be provided during the charrette in a concise and accessible manner (maps, schemes);
4. The drawing is a contract: All drawings produced during the charrette embody the consensus as experienced and achieved by the charrette team. They form a well-understood agreement, or contract, in images amongst the group. The drawings cannot be broken without consent of the group and function as such as a very strong commitment.

7.4.2 The Groningen Charrettes

A specific charrette approach has been developed for Groningen province, in the context of the Hotspot Climate-proof Groningen project (Roggema 2009a). Instead of executing a weeklong or multiple-day charrette, the process consisted of several separated charrettes, lasting for a day. Each of the charrettes were organised separately, but were well-connected through the participation of s small group in every charrette. After the kick-off, several thematic charrettes were held, in which a specific group of experts visualised their optimal climate adaptation future for the specific theme. The themes were the coast, agriculture, water management, nature, water supply and energy. After these series the results were collected and documented on maps. These individual maps were subsequently integrated in one climate adaptation map for Groningen province. Meanwhile two design charrettes were conducted to develop integrated future scenarios. The participants in these charrettes consisted of a mix of people: experienced policy-makers, experts and generalists, in combination with students and people with local knowledge. This group used coloured clay to visualise their optimal future (Fig. 7.9).

The main outcome out of these exercises were a set of integrated future scenario's, the so-called 'wishing-cards', representing the desires of the participants to best make Groningen climate-proof (Fig. 7.10). These scenarios formed the background of possible long-term futures, against which the climate adaptation map could be judged.

This judgement brought several tension areas to the fore, specific areas for which the climate adaptation map was not satisfying in one or more of the scenario's. These areas were designed in more detail in an ultimate Design Charrette and further integrated in two climate proof perspectives: Give Up, in which the potential

Fig. 7.9 Policy-makers in action during the Groningen Charrettes

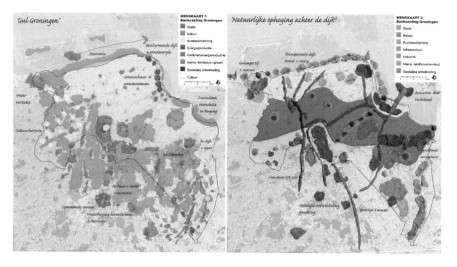

Fig. 7.10 Working with coloured clay: two 'wishing-cards' in the Groningen Charrettes

threat of a flood urges the people to retreat, and Sustain, in which a 'superdike' of 200 m width protects all assets, making it possible to adapt every function to extreme changes in climate (Fig. 7.11).

A couple of specific decisions regarding the process were made. In the first place all raw material that resulted from each individual charrette, was redrawn in precise and beautiful maps. A mapmaker was put in place to produce all this important work. Secondly, for each of the charrettes a specific venue was chosen. The water supply charrette was held in a historic rural estate (a '*borg*'), the agricultural charrette on a farm and the coastal defence one in a hotel on the outside of the dike (Fig. 7.12).

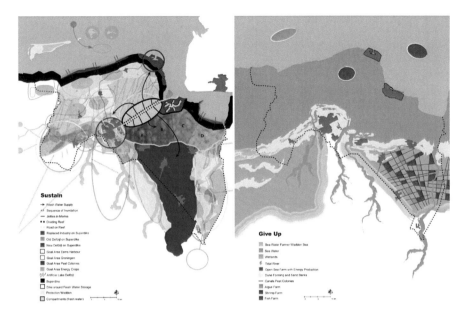

Fig. 7.11 Final result of the Groningen Charrettes: sustain and give up scenarios (Roggema 2009a)

Fig. 7.12 The venue of the 'Coastal Defence' Charrette: Delfzijl seaside hotel

7.4.3 The Victorian Design Charrettes

The design charrette approach, used in three case studies in Victoria, Australia, indicates through action research how people can connect with their future climate and together design future landscapes. The design charrettes in Victoria were

Fig. 7.13 Experts in action during the Bendigo Design Charrette

specifically tailored for the specific context they're held in (Roggema et al. 2010). The method used in Victoria consists of five phases: preparation, design charette 1 (visioning), appraisal, design charrette 2 (implementing) and reporting. After the preparation, in the first design charrette future scenarios for climate adaptation are developed. These scenarios must be seen as charcoal sketches of the future: abstract in scale, but clear in content. The results of the first charrette are then appraised. The quality of the proposed scenarios in terms of their environmental, economic and social value is assessed and this information is used in the second charrette. This charrette aims to design the region in a more detailed way, defines strategic catalyst projects and formulates an investment strategy. This design-assess-design approach is used in three regions: City of Greater Bendigo (a central Victorian major centre, vulnerable for bushfires), and the town of Sea Lake in Buloke Shire (a farming community under threat of droughts, heat and occasional heavy rain).

Both design charrettes are shaped in a very intensive and efficient 2-day meeting. The reason behind this is, apart from the time constraints of individual participants, to create a true intense and highly dynamic session, in which people are more committed and 'into it'. The fact that in regional communities, many people already know each other makes a process of getting to know each other in these cases superfluous.

The standard Victorian charrette program consists of the following key elements: an introductory session in which the urgency of the assignment becomes clear, several design sessions, each of different character, intermediate internal presentations (Fig. 7.13), final design session and presentation of results to an executive panel. In the final design session the optimal climate design is visualised by the participants making use of plasticine (Fig. 7.14), which not only makes it fun to work with but it gives also a 3D-dimension to the work. Each of the separate sessions is highly visual, makes use of mapping or other visual techniques to ensure creativity and thinking beyond the 'window of no'.

During the charrettes, people collaboratively co-design on several levels in attempts to achieve and formulate responses to a complex climate adaptation problem, develop landscape design concepts that respond to future climate change, shape icons that bind, shape and construct models using tacit tools and shape relationships that last.

7 Swarm Planning Methodology 159

Fig. 7.14 Results of working with plasticine; Bendigo Charrette

7.4.4 Key Success Factors

Experiences with both design charrette processes in the Netherlands and Australia reveal the following success factors:

1. *Deal with complex issues*: The first, and maybe most important advantage of working in a design charrette, is that it is possible to make complex issues, such as climate adaptation concrete and conceivable. In the context of a design charrette people easily become creative and will come up with proposals that go beyond the expected and accepted.
2. *Give reason*: The second element of success is the way the results of the charrette are linked and becoming part of regular planning projects after the charrette is finished. In order to enhance this linkage support of the responsible people within the government is essential. When these people make the importance of the subject dealt with in the charrette clear, the sense of urgency in the entire organisation will be felt.
3. *Atmosphere*: The sphere in which the charrette takes place is important, because in a relaxed, but serious environment people tend to perform best. The atmosphere created is one of intensive work and creativity, working towards end results and presentations, and enjoying the work. Benefit of this atmosphere is that boundaries between organisations and people will drop over the course of the charrette. Where in regular circumstances relations are often based on power and interests, the charrette environment provides an atmosphere to engage in a positive discussion on the basis of expertise and the content. The value to step outside routines and planned behaviours allows

participants to connect on a personal level with each other and their environment.
4. *The venue*: When the venues of the design charrettes are deliberately chosen, are special and well connected to the topic, they force participants to step out of their normal routines and open their minds for a creative and intensive session. Participants more easily develop a joint feel of collaboration.
5. *Visual techniques*: The use of maps, clay and plasticine offer an easy way to capture ideas that otherwise might have been forgotten. Moreover, it challenges people to open their minds to new ideas. These techniques stimulate people to use their left, creative and intuitive brain-side. This opens the way to creation of new joint visions on the desired future.
6. *Engage people*. The right mix of people to participate in the charrette depends on the attractiveness of the charrette. There must be a serious reason why it is interesting for people to take part. Apart from this the program must be short, intensive and offer opportunities to network. Key factors in composition of the participants are representation of a mix of experiences, background, ages and places, involve people with an open mind and certainly not to limit the group to only designers. If, for whatever reason, this combination of people could not be secured, it is better to cancel the charrette.
7. *Celebration*: At the end of a charrette the results must be celebrated. The focus in the celebration lay on illuminating the new and exiting ideas that had came up during the process and the fact that all participants could be proud of the achievements made during the process.

7.5 Swarm Planning Experiment

During the Scientific conference 'Smart and Sustainable Built Environments (SASBE)', which has been held in June 2009 in Delft [www.sasbe09.com], a special session was organised to explore the concept of Swarm Planning (Roggema 2008a, b, 2009b; Roggema and Dobbelsteen 2008). A group of approximately 30 scientists were invited to take part in the event and apply the Swarm Planning concept to the Eemsdelta region in the Netherlands. The session consisted of short introductions by several experts in the field and, for the major part, of a *'Living Lab'*, examining the concept and using it in a case study. A Living Lab is a user-driven open innovation (eco)system, which enables users (in this example: the scientists) to take an active part in the research, development and innovation process (European Commission 2009; Pallot 2009; http://livinglabs.mit.edu).

Central rules of the game for a creative thinking session are: postpone judgements, be open within the group and obey privacy outside it, and be modest and build on and enrich the ideas of others. The process consists of a couple of subsequent diverging and converging phases, which, as time passes, come closer to conceptualisation and realisation (Fig. 7.15).

7 Swarm Planning Methodology

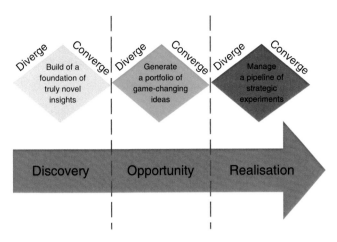

Fig. 7.15 Phases of subsequent divergence and convergence (Van Haren and Starmann 2009)

Fig. 7.16 Scientists at work during the 'Swarm Planning Experiment', SASBE-Conference, Delft

The framework of the Swarm Planning Experiment consisted of several elements, mainly derived from the COCD-box methodology [www.cocd.org/nl/node/53].

The sessions started with four short presentations in which the scene was set. The talks (Geldof 2009; Timmermans 2009; Foliente 2009; Roggema 2009c) informed inspired and opened the minds of the attendees for creativity. After these introductions the first diverging phase of collecting ideas started. In this first step of the process, four groups of participants (Fig. 7.16) tried, in 7 min, to come up with as many ideas as possible, answering four questions. In a second round, the next group built on the ideas of the first group, by adding as many new ideas as possible. This process was repeated until every group had the opportunity to add ideas to each of the four sub-questions.

In the next (converging) step ideas were selected. The tool used during this step was the COCD-box [www.cocd.org/nl/node/53; http://newshoestoday.com]. This tool was developed by the COCD (Centre for Development of Creative Thinking). At the cradle of every paradigm shift there stands an impossible or unsuitable idea. The COCD-box helps to prevent falling into the Crea-Dox: The moment someone thinks

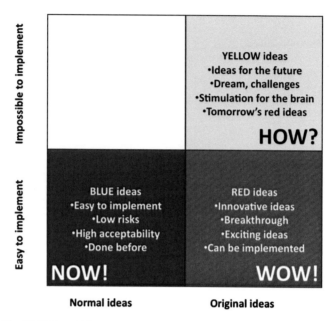

Fig. 7.17 The COCD-box (After: www.cocd.org)

of nice new ideas, he/she remains with the old (known) solutions. To prevent that, a classification can be made using a matrix: the COCD-box (Fig. 7.17). The box has two axes: the originality of ideas and the ease of implementation of ideas. Original and not (yet) feasible ideas are placed in the yellow square, original and feasible in the red square, and feasible and already known ideas are put in the blue square. The box makes it possible to subdivide all ideas and not to lose any idea, something that happens often in regular processes in which especially the yellow ones are easily forgotten. The participants, using yellow, red and blue stickers, select the best ideas in every category by putting the coloured stickers on the ideas of their liking. After this has happened an overview over the best ideas, as seen by the participants, emerges.

The next step in this process is to create concepts out of the selected ideas (Fig. 7.18). The most valued red ideas are put in the upper-left square of the COCD-box and are subsequently enriched with both *dreams* (yellow ideas that strengthen the red idea) and *quick-wins* (the blue ideas that strengthen the red idea). The combination of original ideas with dreams and quick-wins leads to a set of ideas that can be conceptualised into one comprehensive concept for every red idea.

This step is followed by the application of the developed concepts in a spatial design. Originally, this step was not part of the COCD method. Every group takes up one concept and develops a design for this concept on a topographical map of the case study area. This results in four spatial distinct concepts of Swarm Planning for the Eemsdelta region. The results are discussed in Chap. 8. The Swarm Planning Experiment ends with flash presentations by the four groups, illuminating their newly designed concepts (Fig. 7.19).

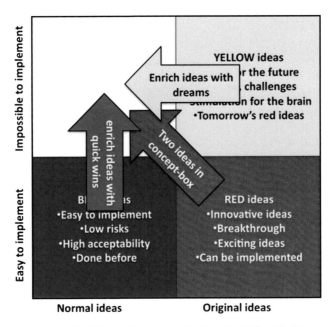

Fig. 7.18 Enriching the original ideas with dreams and quick wins (After: Van Haren and Starmann 2009)

Fig. 7.19 Presenting the results: Wim Timmermans and Greg Keeffe (*left*) and Nils Larsson and Chrisna du Plessis (*right*)

7.6 Conclusion

The Swarm Planning concept is a complex theory. Therefore, the development of a methodology, which makes the application of the concept possible in a practical way, is important. Swarm Planning theory leans heavily on complex adaptive systems theory and focuses on linking time dynamics, or the pace things change, with specific spatial elements in the landscape. The layer approach, which attaches spatial elements to several time-rhythms, is an excellent way to achieve the connection. In this chapter the development of a Five-Layer Framework has been developed, which makes it possible to connect every spatial element to one of the layers. The five layers, networks (1), focal points (2), unplanned space (3), resources (4) and occupation patterns (5), can also be used as a methodology to create a climate adaptive design.

The use of a Swarm Planning Framework requires a specific process. Both the creative thinking, used in the Swarm Planning Experiment as well as the design charrettes offer valuable and useful process methodologies, within which the creativity of participants is triggered and the required swarming attitude can be further developed. Especially working with visual techniques, in the form of maps, drawings and the use of clay and plasticine, opens up the left brain-half and allows people to contribute innovative solutions, which normally stay out of reach.

In dealing with the uncertain and wicked issue of climate adaptation these approaches are essential, because regular planning processes are build on former experiences, limited through existing policies and dominated by organisational interests. The issue of climate change adaptation is too pressing to let old-fashioned habits, procedures and solutions stand in the way of an innovative climate-proof future.

References

Adger WN (2003) Social capital, collective action, and adaptation to climate change. Econ Geogr 79(4):387–404

Commonwealth of Australia (2007) Tackling wicked problems; a public policy perspective. Australian Government/Australian Public Service Commission, Canberra

Condon PM (2008) Design Charettes for sustainable communities. Island Press, Washington, DC

Conklin J (2001) Wicked problems and social complexity (p 11); CogNexus Institute. [online]: http://cognexus.org/wpf/wickedproblems.pdf, visited 13 Dec 2010

De Hoog M, Sijmons DF, Verschuuren S (1998) Laagland. Eindrapportage HMD werkgroep Herontwerp. Gemeente Amsterdam, Amsterdam

De Jonge (2009) Landscape architecture between politics and science. An integrative perspective on landscape planning and design in the network society. PhD thesis Wageningen University and Research Centre. Uitgeverij Blauwdruk/Techne Press, Wageningen

European Commission (2009) Living labs for user-driven open innovation, an overview of the living labs, methodology, activities and achievements. European Commission, Brussels

Foliente G (2009) Urban sustainability transition & adaptation – frontier challenges in complexity. Presentation during SASBE09 special session. SASBE09, Delft

Frieling DH, Hofland HJH, Brouwer J, Salet W, de Jong T, de Hoog M, Sijmons D, Verschuuren S, Saris J, Teisman GR, Marquard A (1998) Het metropolitane debat. Toth Uitgeveij, Bussum

García B (2005) De-constructing the city of culture: the long term cultural legacies of Glasgow 1990. Rev Issue Urban Stud 42(5/6):1–28

Geels FW (2002) Technological transitions as evolutionary reconfiguration processes: a multilevel perspective and a case study. Res Policy 31:1257–1274

Geels FW (2005) Processes and patterns in transitions and system innovations: refining the co-evolutionary multi-level perspective. Technol Forecast Soc Chang 72:681–696

Geels FW (2011) The multi-level perspective on sustainability transitions: responses to seven criticisms. Environ Innov Soc Trans 1:24–40

Geldof G (2009) Climate change and complexity. Presentation during SASBE09 special session. SASBE09, Delft

Jones R (2011) Planning with plasticine. http://2risk.wordpress.com/2011/11/30/planning-with-plasticine/. Accessed 30 Nov 2011

Lennertz B, Lutzenhiser A (2006) The Charrette handbook. The essential guide for accelerated collaborative community planning. The American Planning Association, Chicago

Nonaka I (1994) A dynamic theory of organizational knowledge creation. Organ Sci 5(1):14–37

Nyseth T, Pløger J, Holm T (2010) Planning beyond the horizon: The Tromsø experiment. Plan Theory 9(3):223–247

Palot M (2009) Engaging users into research and innovation: the living lab approach as a user centred open innovation ecosystem. Webergence Blog. http://www.cwe-prpojects.eu/bscw.cgi/1760838?id=715404_1760838

Polanyi M (1966) The tacit dimension. Routledge & Kegan Paul, London

Portugali J (2000) Self-organisation and the city. Springer, Berlin/Heidelberg/New York

Rittel H, Webber M (1973) Dilemmas in a general theory of planning. Policy Sci 4:155–169. Elsevier Scientific Publishing Company, Inc., Amsterdam [reprinted in Cross N (ed) (1984) Developments in design methodology. Wiley, Chichester, pp 135–144]

Roggema R (2008a) The use of spatial planning to increase the resilience for future turbulence in the spatial system of the Groningen region to deal with climate change. In: Proceedings UKSS-conference, Oxford

Roggema R (2008b) Swarm planning: a new design paradigm dealing with long-term problems associated with turbulence. In: Ramirez R, Selsky JW, van der Heijden K (eds) Business planning for Turbulent Times, new methods for applying scenarios. Earthscan, London/Washington, DC, pp 103–129

Roggema R (2009a) DESIGN, Hotspot climate proof Groningen, Final report. Province of Groningen and Climate Changes Spatial Planning, Groningen

Roggema R (2009b) Adaptation to climate change, does spatial planning help? Swarm planning does! In: Brebbia J, Tiezzi S (eds) Management of natural resources, sustainable development and ecological hazards. WIT Press, Southampton, pp 161–172

Roggema R (2009c) Swarm planning principles. Presentation during SASBE09 special session. SASBE09, Delft

Roggema R (2012) Developing a planning theory for wicked problems: swarm planning. In: Van den Dobbelsteen A, Stremke S (eds) Sustainable energy landscapes. CRC/Taylor & Francis, Abingdon

Roggema R, van den Dobbelsteen A (2008) Swarm planning: development of a new planning paradigm, which improves the capacity of regional spatial systems to adapt to climate change. In: Proceedings world sustainable building conference (SB08), Melbourne

Roggema R, Horne R, Martin J (2010) Design-led decision support for regional climate adaptation; VCCCAR research proposal. VCCCAR, Melbourne

Roggema R, van den Dobbelsteen A, Biggs C, Timmermans W (2011a) Planning for climate change or: how wicked problems shape the new paradigm of swarm planning. In: Conference proceedings, 3rd world planning schools congress, Perth, 4–8 July 2011

Roggema R, Martin J, Horne R (2011b) Sharing the climate adaptive dream: the benefits of the charrette approach. In: Refereed proceedings of the 35th annual conference of the Australian and New Zealand regional science association international, Canberra, 6–9 Dec 2011. Edited by Paul Dalziel AERU research unit, Lincoln University, Lincoln, New Zealand

Roggema R, Van den Dobbelsteen A, Kabat P (2012) Towards a spatial planning framework for climate adaptation. SASBE 1(1):29–58

Timmermans W (2009) How to understand planning by surprise? Presentation during SASBE09 special session. SASBE09, Delft

Van Haren R, Starmann I (2009) Creative problem solving for wicked problems and swarm planning. Presentation during SASBE09 special session. SASBE09, Delft

VROM-raad (2007) De hype voorbij, klimaatverandering als structureel ruimtelijk vraagstuk. Advies 060. VROM-raad, Den Haag

Websites

http://livinglabs.mit.edu. Accessed 14 Dec 2010
http://newshoestoday.com
www.cocd.org/nl/node/53. Accessed 14 Dec 2010
www.sasbe09.com
www.mt.unisi.ch/pages/barcelona.pdf. Urban Transformation and 92' Olympic Games, Barcelona. Accessed 30 Jan 2012

Chapter 8
Swarming Landscapes

Rob Roggema

Contents

8.1	Introduction	168
8.2	Strategies	168
	8.2.1 Impulses	168
	8.2.2 Steer the Swarm	170
8.3	Interventions	171
	8.3.1 Groningen Museum	172
	8.3.2 Blauwe Stad	172
8.4	Climate Landscapes	174
	8.4.1 Floodable Landscape	174
	8.4.2 Bushfire Resilient Landscape of Murrindindi	177
	8.4.3 Bushfire Resilient Landscape of Bendigo	183
8.5	Swarm Planning Experiment	187
	8.5.1 Time Incongruence	188
	8.5.2 On the Move	189
	8.5.3 Sustainable Emergence	190
	8.5.4 Destructive Mob-Elections	191
8.6	Conclusion	192
References		192

Abstract The Swarm Planning Theory and Methodology, as outlined in Chaps. 6 and 7 have been used in several design projects in the recent past. Swarm Planning can be applied in different ways. It can be used to formulate a spatial strategy, as the examples of Strategic interventions and steer the swarm show. It can also be used to identify the location and the type of intervention to be taken. This is illustrated through the Groninger Museum and Blauwe Stad examples. The third way of applying Swarm Planning is to design climate landscapes, in which not only the intervention

R. Roggema (✉)
The Swinburne Institute for Social Research, Swinburne University of Technology,
PO Box 218, Hawthorn, VIC 3122, Australia
e-mail: rob@cittaideale.eu

is identified but also the dynamic impact in the landscape is part of the design. The examples of a Floodable landscape and the two Bushfire resilient landscapes of Murrindindi and Bendigo illuminate this. Finally, Swarm Planning can be used to create innovative spatial solutions for a specific design assignment. The laboratory setting of the Swarm Planning Experiment proves the innovative capacity of the approach. All four applications are seen worthwhile and support the potential of the Swarm Planning theory, methodology and use.

Keywords Swarming landscapes • Interventions • Strategies • Climate landscapes • Swarm experiment

8.1 Introduction

Each site is different and the design for every location will be different. The benefit of using the Swarm Planning methodology is that dynamic character of spatial elements will be included in the design and specific interventions, based on the analysis of networks, can be determined and provide resilience in the area to anticipate future climate impacts. In this chapter the results of various Swarm Planning designs are presented. The first section describes a couple of Swarm strategies (Sect. 8.2), while in the second section (Sect. 8.3) specific interventions are highlighted. The design of climate landscapes forms the content of Sect. 8.4 and the chapter ends with presenting the results of the Swarm Planning Experiment.

8.2 Strategies

The result of Swarm Planning eventually takes the shape of a concrete spatial design. The Swarm Planning concept however urges also to think about the underlying strategies. Instead of reacting to a spatial design problem, Swarm Planning anticipates the future and tries to direct the future in a direction that is better capable of dealing with the impacts of climate change. This means that interventions in the spatial domain need to be taken before a climate threat or hazard can be expected. This makes Swarm Planning extremely strategic. In this section two strategies are illuminated: 'strategic impulses' and 'steer the swarm'.

8.2.1 Impulses

To improve preparation for turbulent environments, such as specific impacts of climate change strategic interventions need to be put in place. These interventions are not meant to define in detail the final state an area is planned to become, but they mark the start of a process, which allows emergence of the area on its own from the moment in time the intervention is executed and will start influencing a larger area. To a certain extent, the effect of an intervention may be predicted. The intervention needs to make plausible to be generating increased resilience in the area, but the exact future spatial shape the area will guise is impossible to define. The projection of interventions instead of executing detailed spatial plans makes it possible for stakeholders, involved parties and citizens to co-operate and contribute to the development of the area, because not all is decided on and 'cast in cement'. The increase of resilience is realised by loosening the fixed planning system of the existing. In many occasions a fixed status is the cause of large risks. Introducing flexibility to deal with future uncertainties, threats and challenges the resilience can be enhanced. When space is created for the impact of these threats and developments might include, society is better prepared for future events. Moreover, these interventions in the spatial system introduce the possible impact at a slow pace, which makes it possible for inhabitants to get used to a situation that will be the new normal on the longer term. The *'windows of Groningen'* (Fig. 8.1) show several of these opportunities (Roggema 2008), where loosening the tight and normative planning rules enable the area to react proactively and to increase preparedness.

These strategic interventions are the impulses, which are added to the area and adjusting the area without changing the functions. The impulses make use of the capacity the spatial system has to adjust itself to new circumstances and developments. In the Groningen case several of these interventions have been identified (Fig. 8.1). They have in common that a single intervention opens the way to an indirect effect in a larger area.

1. Heightening the closure dam of the Lauwers Lake enables the area to store more rainwater in winter, influencing the entire catchment area of the Reitdiep-river;
2. Creating new kwelderworks near the Eems harbour enables the Wadden Sea to create new arable land through natural accretion of sand and clay particles. This new land may be used according future demands: as agricultural land, for industrial purposes or as an ecological zone;
3. Perforating the dike between the Eems harbour and Delfzijl opens the opportunity to create a dynamic coastal system, which is able to supply the hinterland with sand, that sedimentates and raises the topography here. The rise of the land happens at the same pace or even faster as the sea level rises. The area can be used and occupied as an innovative living area;
4. Moving the Sea sluice of Delfzijl outside the city makes it possible to create a safer storm surge barrier and offers Delfzijl the chance to develop its waterfront towards the sea;
5. Generation of a luxurious living area around a new lake in a back-dropped area of Eastern Groningen makes it possible to extend the capacity for water storage and has a positive influence on the living standards in the entire area;
6. The introduction of a new railroad, which connects the City of Groningen with the Peat Colonies, enables the southern part of the province to develop a robust ecological corridor, which gives space to shifting ecological habitats and makes an interesting living area possible amidst nature.

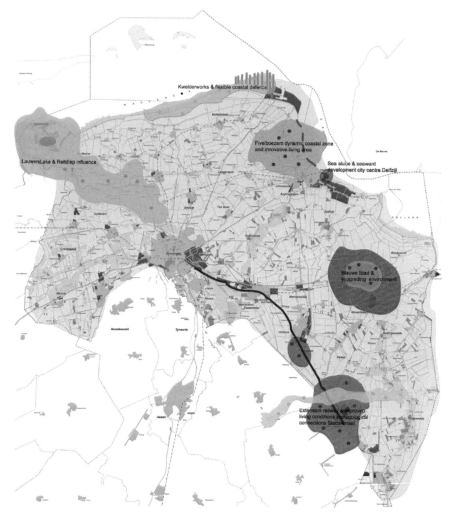

Fig. 8.1 The 'Windows of Groningen', strategic swarm interventions (Roggema 2008)

8.2.2 Steer the Swarm

A slightly different strategy has been developed in preparation for the Regional Plan for Groningen Province. In the analysis (Roggema and Huyink 2007) the so-called steer the swarm strategy has been developed. Existing spatial planning processes have difficulties to create creating effective interventions, which can anticipate future turbulent change, such as the uncertain impacts of climate change and the non-fossil energy supply. These kinds of topics are complex, long-term, uncertain and will become manifest in the far future. In traditional terms: they cannot be planned, but will occur as surprises. It is possible to deal with these uncertainties when areas are given the opportunity and the space to adjust anticipative

Fig. 8.2 A 'Mondrian' combination of interventions (*left column*) and identities (*right field*) (Roggema and Huyink 2007)

or simultaneously with sudden changes. By starting interventions immediately, experience for future circumstances, threats and challenges, can be built up straight away. The resilience in the area is enhanced when space is created in the spatial lay-out and in plans for unforeseen demands in the form of buffers.

In order to achieve this, the planning system of the future needs to include steering principles, enabling areas to adapt more easily and change its spatial patterns as required. These steering principles include two elements: space to change and strategic interventions. These elements differ from place to place, depending on the characteristics of the natural system and spatial identity. Thus, the swarm needs to be steered accordingly.

Depending on the specific qualities of the area the spatial regime can be distinguished. Each specific identity, consisting of the natural (horizontal axis) and spatial (vertical axis) character of the system (the coloured patterns in the right hand side of the black bar, Fig. 8.2) is directed by a combination of well-defined interventions (the left column, Fig. 8.2). Hence, creating a specific resilience regime (e.g. the specific colour for a certain area). In abstract terms, a 'Mondrian' typology arises (Fig. 8.2) as has been developed in the draft 'Atlas Groningen' (Roggema and Huyink 2007). The swarm is steered in a tailored way, intervention x identity dependent.

8.3 Interventions

Core of the Swarm Planning theory is that a spatial system, seen as a complex adaptive system can be steered and stimulated to perform resiliency or higher levels of adaptation. Crucial factor in the theory is it estimates this happens as a result of well-identified and placed interventions. These types of interventions are not new,

 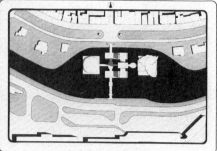

Fig. 8.3 The Groninger Museum positioned in the Verbindingskanaalzone, before (*left*) and after (*right*) the intervention

nor only useful in the case of climate change. In this section two examples from recent history are described: the Groninger Museum and Blauwe Stad.

8.3.1 Groningen Museum

The 'Verbindingskanaal' (connection channel) is a waterway at the edge of the city-centre of Groningen, located between the central station and the inner city (Fig. 8.3). In the past, the only way to reach the inner city was to walk around the canal. As a consequence the area at the city side of the Verbindingskanaal became neglected, attracted hooligans and criminals. On a day, a councillor of the municipality decided on to build the cities museum in the middle of the canal (Fig. 8.3), connecting the station with the inner city. As a result of this, the neglected part of the city centre changed into a very lively and attractive area, used by a large number of people. The intervention of building the Groninger Museum (Fig. 8.4), exactly at this location, had widespread effects and transformed the entire city.

8.3.2 Blauwe Stad

Fact one: the eastern part of the Province of Groningen has traditionally been one of the poorest regions in the Netherlands with pervasively high unemployment, low levels of education and extraordinary poverty. People, who could, left the area.

Fact two: due to more intense periods of heavy rainfall the need to find water storage in the lowest parts of the province is urgent.

Fact three: The Blauwe Stad area is one of the lowest places in the province and this fact, in combination with the back-dropped character of the area, has led to the decision to build, around a newly dug artificial lake, a new neighbourhood: the Blauwe Stad (Fig. 8.5).

Fig. 8.4 The Groninger Museum (Picture: ©Rob Roggema)

Fig. 8.5 Blauwe Stad implemented in the landscape of eastern Groningen

This intervention has multiple effects. It results in the upgrading of the entire area, led to economic growth, increase of amount and quality of amenities, improved infrastructure, increased the possibility to deal with large amounts of water and decreased unemployment. The area evolves by itself after the impulse of the Blauwe Stad has been given.

8.4 Climate Landscapes

Interventions can lead to unpredictable developments in the future, but when the Swarm Planning process is taken one step further the future spatial adjustments can be estimated. And, in reverse, when a certain change is desired in the landscape, the right intervention can be invented in order to meet those demands. In this section three designs for these climate landscapes are described: the floodable Landscape and the Bushfire resilient landscapes of Murrindindi and Bendigo.

8.4.1 Floodable Landscape

The Eemsdelta region is located in the northeast of the Netherlands and consists of two industrial harbours and a valuable heritage hinterland, where remains of old artificial hills, the so-called *wierden* and historic churches dominate the landscape. Despite the economic activities in the area and the growth of jobs, the population is shrinking, putting amenities in and liveability of villages under pressure. The area is not very popular to move to and many (younger) people leave the area. The area confronts the Wadden Sea area and dikes defend its coast. However, current coastal defence standards are not met. The weakest point in the coastal defence of Northern Netherlands lies in this area, which places the area for which sea level rise is expected under threat of flooding when a spring tide occurs in combination with a severe storm. An eventual flood will reach the capital city within 36 h and threatens the national gas reserve, which is also found in the area. There are slight topographical differences apparent in the area of around 1.5 m. The highest areas are found closest to the coast. Sea level potentially rises to 1.3 m above current level and relative sea level rises even more because of maximal 40 cm of soil-subsidence as result of gas extraction. The expected change in wind patterns, turning to the northwest will increase the surge of water towards this region, resulting in higher risks at severe storm surges and potential flooding (Fig. 8.6).

In the current discourse in dealing with coastal defences for sea level rise and storm surges, the safety level is increased through the strengthening and heightening of protecting structures, such as levees and dikes. Fast and accelerated sea level rise as predicted by Hansen and others (Hansen 2007; Hansen et al. 2007, 2008; Hamilton and Kaiser 2009; Lenton et al. 2008; Rahmstorf et al. 2007; Tin 2008) give reason to concerns over the capability of defences to withstand extreme circumstances at

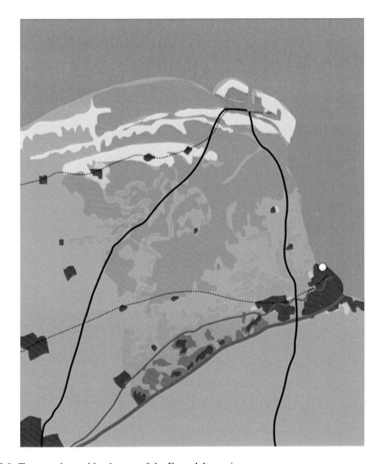

Fig. 8.6 Topography and landscape of the Eemsdelta region

all times. Eventually, even the strongest dike will breech. The consequence of this belief in defending the values behind an increasingly stronger dike is that once it breeches these values are highly vulnerable for the effects and a huge disaster destroys most of the values, such as properties, productive land or human life.

Given the uncertain pace of sea level rise and the moment a dike eventually will breech, the question can be raised if alternative designs may potentially be better equipped for decreasing the impact of sea level rise and storm surges.

The main driving climate forces in the Eemsdelta design are sea level rise, storm surge and the topography of the hinterland. These three forces determine the places and the level of a potential flood. Imagining a potential dike breech at a certain location at a certain sea level, the parts of the landscape that will flood can be determined in detail. When different sea level rise scenarios are taken into account the different stages of flooding can be estimated and a future occupation pattern can be designed able to deal with every possible flood level.

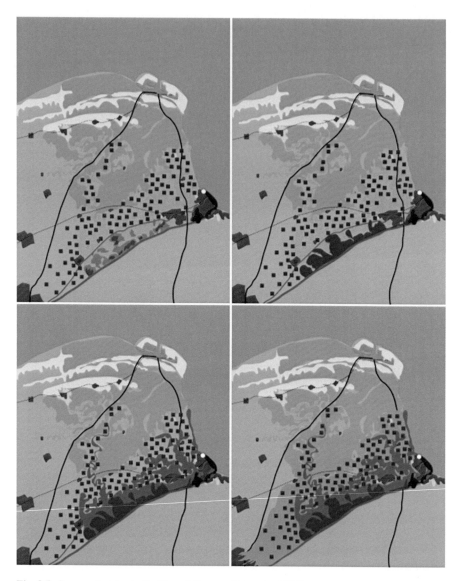

Fig. 8.7 Swarm planning in the Eemsdelta area (Roggema 2009)

8.4.1.1 Intervention at the 'Whole'-System Level

The crucial factor in the design for the Eemsdelta area (Fig. 8.7) (Roggema 2009) is the changed viewpoint from which the problem is approached. Instead of trying to increase the protecting level through strengthening structures, an advanced crucial intervention is proposed, which influences the system as a whole. Instead of keeping water out, it is let in at a very slow pace and in a very well predictable way.

A hole in the coastal defence allows water to enter the hinterland and the level will rise as sea level rise increases. The analysis of networks determines the most crucial and also most vulnerable nodes in the coastal defence. Here (the white dot in the maps, Fig. 8.7), the focal point for a strategic intervention is found. This is the point where water is allowed to enter the hinterland.

8.4.1.2 Freedom to Emerge

As a result of the crucial intervention, allowing a hole in the dike, all individual elements are challenged to perform emergent behaviour. Behind the intervention point, the whole in the dike, the 'unplanned' area, which is kept free from any function, is identified. For every single rise in sea level a different area is required for inundating water. Sea levels of 0.3, 0.6, 0.9 and 1.2 m, the latter being the highest level estimated for 2100 by the Dutch Delta Committee (Deltacommissie 2008) determine where it is safe to live and where adaptation is required. Specific areas are reserved for the storage of water, the production of sustainable energy resources and food, all kept outside the maximal inundation area. On the edge between the floodable landscape and higher and dryer places occupation emerges. This twilight zone will face wet circumstances only when sea level rises 1.2 m, but the buildings built in this area can withstand or even profit from being in the middle of sea water. The buildings are constructively adjusted to wet environments (Fig. 8.8), made waterproof, floating (Fig. 8.9) or made suitable to function on both land and in water (amphibious).

The advantages of this design are that impacts as result of a big disaster are prevented, because the water is already allowed in the hinterland and used as an ally and not as an enemy. Because of the fact that accurately can be predicted where the water will flow, people, buildings and organisation are very well capable of adapting at a very early stage. The water will bring gradually changes and benefits. At first, in the unplanned areas brackish conditions emerge, allowing ecological conditions to enrich. Secondly and at a later stage all new buildings face water in their environment, a real estate asset of great value. Probably the biggest advantage is that due to the slow pace of entering seawater a disaster never happens, but it is tamed to a gradually changing wet environment, which makes the area inherent safe. The proposed intervention may help inhabitants in the area to slowly adapt to changes, instead of being surprised by a sudden flood.

8.4.2 Bushfire Resilient Landscape of Murrindindi

The Kinglake-Murrindindi region is located to the North-East of Melbourne Metropolitan area. In the region several small communities and villages, surrounded by agricultural land, are located amidst a hilly and mountainous landscape. Abundant forests and ecologically valuable nature reserves cover most of the area. Divided by

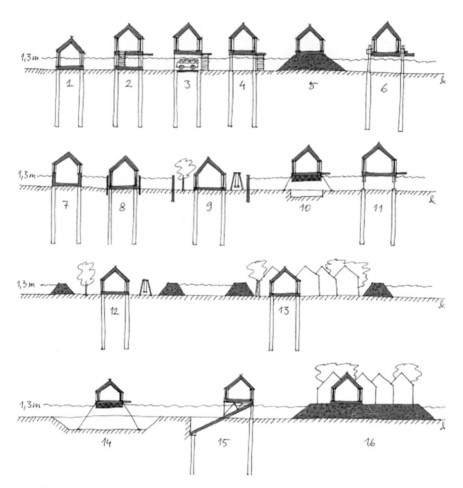

Fig. 8.8 Spectrum of technological innovations to deal with appearance of water in the living environment (TU Delft 2008)

the Great Dividing Range the Northern part of the study area discharges its water to the North (Murray River) and the Southern part to the South (Yarra River) and leaves the area rather quickly. The area is vulnerable for droughts and the warm 'wind tunnel' entering the area during hot summer days from the North. The Kinglake-Murrundindi area is bushfire prone and one of the most severely damaged areas during the Black Saturday bushfires of 2009 (2009 Victorian Bushfire Royal Commission 2009). The area includes both sides of the Great Dividing Range and many types of landscape elements exist: mountainous area, forests, villages, meadows and agricultural use. Despite the fact that the area is very vulnerable for bushfires, people still want to live and recreate there.

The landscape is formed through topographical differences. The Great Dividing Range literally divides the Northern and the Southern areas and functions as a

8 Swarming Landscapes

Fig. 8.9 New forms of living at the edge or on the water (© RAL Architects)

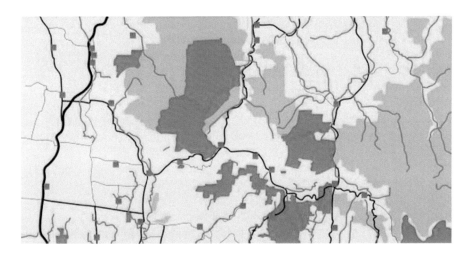

Fig. 8.10 The landscape of the Kinglake-Murrundindi region

watershed (Fig. 8.10). Many of the smaller communities are located at the upper parts of the ridge, where also the main road crosses the area from east to west. The forests are mainly found on the higher grounds and the majority of the water flows undisturbed out of the area. In case of heavy rain floods are likely, but after a period of extensive drought the riverbeds run nearly dry. Many of the (smaller) roads go up or down the hills, but they do not form a circuit.

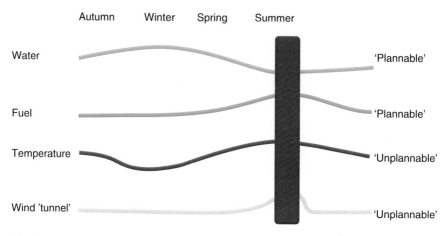

Fig. 8.11 The yearly 'high tide' of risk

The period with highest risk at bushfires is yearly occurring. Most of the year there hardly is any risk. The question how we can make use of this 'no-risk' period to prepare for and anticipate the risk period is interesting as it opens the way to dealing with and planning ahead of a potential disaster, eventually preventing it from happening. The design of the area, which is harvesting water and minimising the amount of fuel in the period before the bushfire season, supports increased safety during this period.

The main drivers to increase the risk at bushfires are (1) increasing temperatures, (2) the acceleration of the 'wind tunnel' effects from the northwest and (3) the amount of fuel (e.g. wood, any plant material that could burn) during the seasons. The combination of these drivers leads to a yearly *'high tide'* of bushfire risk (Fig. 8.11) at the moment of high temperatures, much fuel, hardly any water and a hot and dry wind blowing in from the northwest.

Two of these drivers are 'unplannable' at the scale of the site, because these drivers, the wind-tunnel and the temperature, are exogenous to the area: they originate at the global or continental level, which makes them hard to influence. The amount of water and fuel however, can be influenced at the local level. In the period leading up to the yearly high tide crucial interventions can be taken to minimise risk. When water could be still around and the amount of fuel could be lowest at the moment of the high tide, a landscape can be created in which people can safely live, even in circumstances that would otherwise be dangerous.

8.4.2.1 Intervention at the 'Whole'-System Level

The proposed strategic intervention, which influences the landscape at the 'whole'-system level, is to place small dams in all of the rivers. The exact places of these

8 Swarming Landscapes

dams are identified on the basis of the network analysis of the area and are found at the foot of the mountain-ridge, where low-lying areas are naturally formed. Here, the strategically placed dams can harvest water in naturally created reservoirs. All rainwater that falls year-round is collected and fills up the reservoirs over the subsequent seasons (Fig. 8.12). Starting just after summer the reservoirs are empty and dry and will be filling up in autumn and winter. In spring, when bushfire risk is highest, the reservoirs are filled at the highest levels too. In deciding which reservoirs collect water their positioning is important. A North Westerly wind is the one and only dangerous hot wind, causing devastating bushfires. Positioning the reservoir to the North Western side of existing forests creates an extra buffer and makes living here the safest. Based on the network analysis most of the naturally formed reservoirs are located in these places.

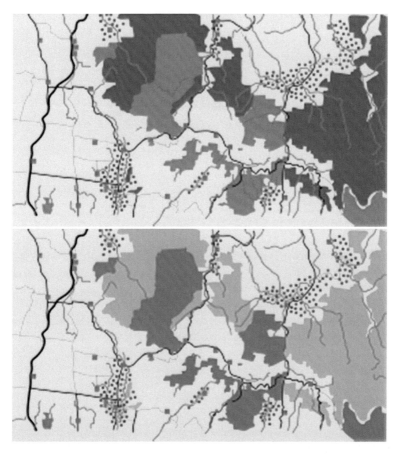

Fig. 8.12 Seasons: the end of summer, the beginning of winter, spring and just before summer

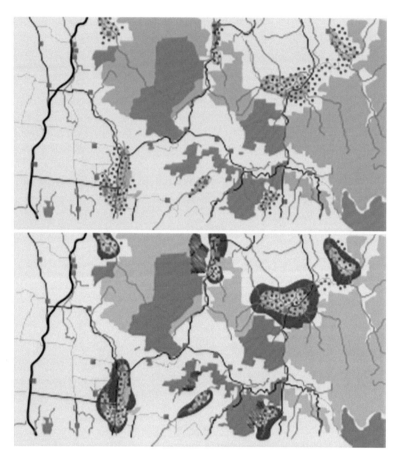

Fig. 8.12 (continued)

8.4.2.2 Freedom to Emerge

Once the interventions at the system level are defined the landscape will take shape as from that point on. The areas that form the reservoirs are dependent on the rainfall; the amount will determine the extent of the reservoirs. In the areas around the reservoirs 'unplanned' space needs to be created, where water can flow is necessary. Around these dynamic reservoirs forests, meadows, agriculture and housing are planned. Any additional forest are planned at the South Eastern side of the reservoirs, agriculture at the edge of the highest expected water levels and meadows are planned where temporarily water can be expected. In and around the reservoirs additional housing can be build, which need to be adjustable to both floods and fires, which leads to the somewhat contradictory proposition of building floodable housing in fire prone areas. These houses are built in what will be temporarily reservoirs over the year, they are placed at distance from potential fires and the water of the reservoirs can be used in an emergency to extinguish the fire. There are not many examples of

8 Swarming Landscapes

Fig. 8.13 Water pavilion (Ties Rijcken)

Fig. 8.14 Bushfire resilient house (John King Architects)

houses that are resilient for water and fire at the same time. Houses are prepared to float, such as the water pavilion (Fig. 8.13) or to be bushfire resilient (Fig. 8.14).

Only recently, solutions for multiple climate hazards are integrated in one design, such as the winning design in the Australian Insurance competition (Fig. 8.15), able to deal with hail, bushfire, heavy rain, flood and cyclones.

8.4.3 Bushfire Resilient Landscape of Bendigo

Swarm Planning theory has been implicitly used for the design of a bushfire resilient landscape in the Bendigo area. This town, in central Victoria, Australia is surrounded by forest and thus extremely bushfire prone. The town was, amongst several other places in Victoria, hit by the bushfires on Black Saturday, 7 February 2009. Bendigo

Fig. 8.15 Bluescope – resilient house, winner of the Insurance competition (Caroline Pidcock)

is one of the fastest growing regional towns in Victoria. It will need to build approximately 23,000 new houses until 2050. At the same time the town is under an increasing threat of bushfires, because (1) the town is surrounded by forests, (2) it is inevitable that new developments will in one way or another enter the surrounding landscape (a push outward) and (3) climate change will exaggerate in the number of hot days and in average high temperatures, leading to an intensified bushfire hazard. In the landscape design for the area (Newman et al. 2011) Swarm Planning interventions are proposed for the entire system (the whole of Bendigo and surroundings) as well as for specific development locations and individual (landscape) elements.

8.4.3.1 Intervention at the 'Whole'-System Level

The major proposition that intervenes in the future development of the entire system is the 'rule' that when a house is destroyed by a bushfire, the house cannot be rebuild but will be replaced by a huge concrete pillar. This pillar symbolises the vulnerability of the place where the house is lost and makes it manifest that the lost house wasn't the most resilient one in the area. Over time only the houses that are best prepared to deal with fire will remain in the area. Because of the fact that the bushfires in this area always originate from the northwest (due to the, on hot days, prevailing hot wind from the central Australian desert), the North Western urban fringe will slowly transform in a zone consisting of the most resilient houses with an increasing number of pillars (Fig. 8.16).

After a while, the North Western zone will start to function as a shield (Fig. 8.17). The combination and positioning of concrete pillars together will protect the

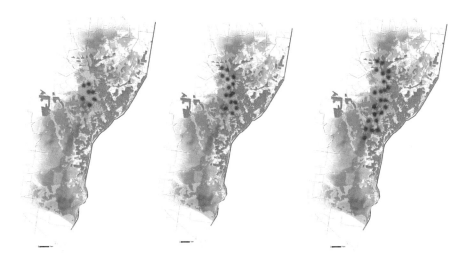

Fig. 8.16 Replacing burnt houses by pillars (the *red dots*); creating safety in the most risky zone (Newman et al. 2011)

Fig. 8.17 Artist impression of the protection zone (Newman et al. 2011)

remaining houses from fire attacks from the northwest, because the shield breaks the wind, and thus preventing the fire to continue its devastating pathway, and it also captures embers, which else would function as the outposts of the fire to start new spot fires in front of the fire-front.

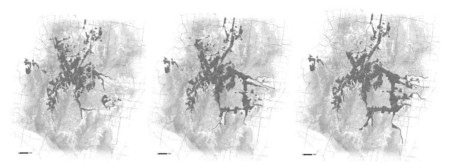

Fig. 8.18 Bendigo 'moves' (Newman et al. 2011)

The total of the urban 'system' will reorganise itself, because the northwest side of town is prevented from new housing, which therefore takes place at the eastern lee side of attacking fires. As result the city slowly 'moves' towards less vulnerable landscapes in the east (Fig. 8.18), meanwhile protected at its most vulnerable North Western side.

8.4.3.2 Freedom to Emerge

The city shape at the eastern side (Fig. 8.19) can be viewed as the level at which individual landscape components in interaction with each other develop emergent properties. The following landscape elements are determined and attributed with complex systems properties:

- Sand-dune: initially the only design intervention that will be realised. The sand is material, which can flow freely in its surroundings, finding shape due to the micro-climatic differences in the specific location. Generally, the sand will form structures according to morphological rules and will ultimately shape as sand-dunes;
- Pillars: once the dunes have formed and reached a more or less stable state the pillars are added in the most strategic spots, namely the places where they have largest sheltering effect. They provide shelter for hot winds from the north and break eventual fire from that direction, but they are also capable of offering shadow, creating a micro-climate where animals and plants can survive during the hottest days;
- Pig face: this is the 'un-burnable' plant, which can be planted in order to prevent fire from progressing. The plants are projected at the bottom of the sand dunes and allowed to grow and expand its territory freely. It will, in interaction with the sand dunes and the pillars find its most optimal places to grow. Because it is initially planted at the northern side of the sand-dunes it will stop grassfires from moving up the sand dune hill;
- Bike path: after the dunes, pillars and pig face have established themselves a bike path is added. The path is projected in a way that it profits optimally from the shelter and shadow the sand dunes, pillars and plants offer;

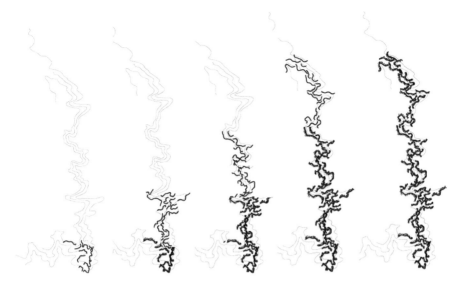

Fig. 8.19 Intervention, followed by occupation in subsequent steps (Newman et al. 2011)

- House: in the last stage of the process people are allowed to build their houses wherever they want to. Because the context is a given, the dunes and pillars created shelter and the bike path accessibility people will generally position their homes (1) behind the shelter, (2) connected to the path and (3) at distance from houses already apparent. This self-organising process will in the end lead to a landscape in which safety, liveability and social responsibility are key values.

Over time the process of growth at the eastern side of town is incremental. As the first dunes are formed and the pillars are under construction, in the next area the dune forming processes can be enhanced. The subsequent process of dune forming, pillar building, planting and occupation leads to a slow occupation and transformation of the landscape (Fig. 8.19), allowing it to adapt to the changing circumstances in an easy way.

The design for Bendigo can be seen as an example of Swarm Planning, in which the core characteristics of complexity are used to intervene in the system and start a process leading to a higher adaptive capacity. The main interventions in this design are the inability to rebuild after burning (and replacement by a concrete pillar) and the initial sand suppletion in the eastern fringe.

8.5 Swarm Planning Experiment

The special session on Swarm Planning for Wicked Problems (see Chap. 7 for background), which took place on 16 June 2009 in Delft, the Netherlands, resulted in four Swarm Planning models for the Eemsdelta region in Groningen: 'time incongruence', 'on the move', 'sustainable emergence' and 'destructive mob-elections'.

8.5.1 Time Incongruence

The specific question this scenario, time incongruence (Fig. 8.20) needed to deal with was how Swarm Planning can be used to increase coherence amongst people and in society. The proposition in this scenario was to focus on creating a common enemy, create a hyper-meeting place and adjust the landscape, settlements and infrastructure networks according time dimensions.

Several incentives are proposed in this scenario. In the first place two dynamics are introduced: the slow, laid back pace of the interior network and the highly dynamic pace of the turbo-meeting-place. These time-incongruencies make it possible to choose your style accordingly, and let it depend on lifestyle or exogenous factors, such as living with or away from climate hazards. The settlements are organised in connected networks and superposition the slow-turbo dichotomy, allowing for the freedom of choice.

The second intervention is to organise a common enemy. In this case this enemy is a storm surge, entering the area from the North. This common enemy is not always reality, but stays for ever as a mental enemy in people's minds, causing disruptive and uncertain environments. The idea to cause this mental disruptions is to introduce an 'award for the worst plan'; not to realise the plan, but to make people aware of what could happen and confront them with 'maybe-planning'. It might,

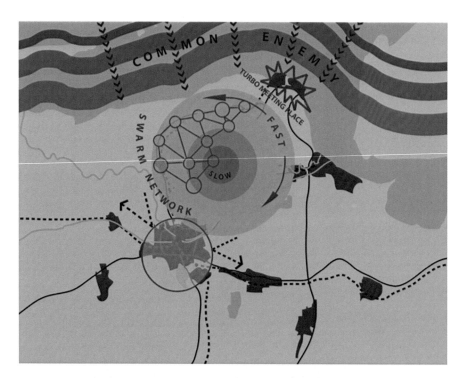

Fig. 8.20 Swarm model one: time incongruence

8 Swarming Landscapes

and it might not happen. The introduction of this uncertainty improves consciousness about peoples' environment, whether it is slow or turbo.

8.5.2 On the Move

The second scenario, on the move (Fig. 8.21) answers the question how to use Swarm Planning in dealing with sea level rise, floods and the fair distribution of fresh water. The proposition in this scenario is to plan a directed flood, occurring in 2059, allowing sea water to flow in the largest part of the region.

Through the introduction of the idea, now, to create a flood, later, two things happen. In the first place is everyone from point zero aware of the fact that in 2059 a flood is going to happen. This flood, then, is no longer a surprise, but a long expected occasion. In anticipation, historic villages can be rebuilt in safer spots, higher and safer villages may get a sister village on the beach and assets can be moved towards the higher landscape in the South. The motto here is 'Dress for the ride, not for the crash', in other words, be prepared to move on time and anticipate, not escape, the future. In the second place this scenario emphasises to accept the powers of nature and climate change. Being conscious about possible future environments

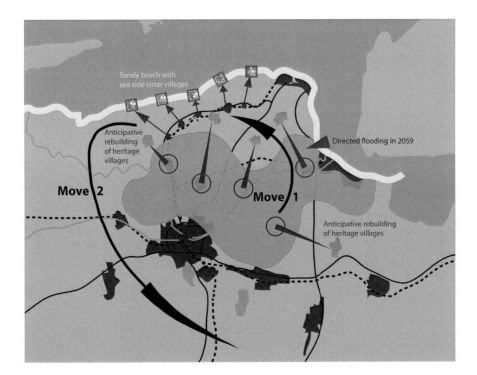

Fig. 8.21 Swarm model two: on the move

allows people to accept (little) floods, start living upstairs and put their furniture on pulleys. The ideas that support this scenario include the creation of unsustainable levees, to remove safety standards and to let structures emerge organically. These small changes in how planning is perceived improves the awareness that humankind is the problem, not climate change.

8.5.3 Sustainable Emergence

In this scenario, sustainable emergence (Fig. 8.22) the question how to improve the identity through the use of Swarm Planning is tackled. The key proposition in this scenario is to abandon planning and design, and to have faith in emergence 'from the ground'.

The scenario emphasises that abandoning planning rules as known to date will enlighten and give freedom to emergence on the basis of natural features of the region. As an outcome this will eventually lead to the development of a new kind of fuzzy rules, which are non-linear (e.g. not for everyone and everywhere the same) and imply all a different kind of incentive, reward or punishment. These rules are not top down anymore, but bottom up, taking the natural, local features as the starting point for designing and let emergence take place from there.

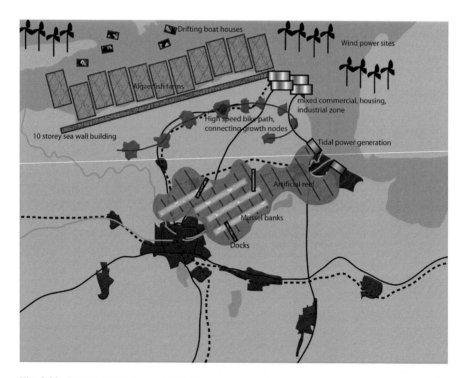

Fig. 8.22 Swarm model three: sustainable emergence

8 Swarming Landscapes 191

A good way to create plans, which understand the local, original qualities, is to listen to people and tell them the truth. Planners need to take a step backwards, start living in their own plans and before entering the planning profession design a chair. This could prevent the implicit drive of planners to build houses everywhere and to let design rules dictate the result. Instead, by taking into account the natural features of the area, an inland sea with mussel-banks can be envisioned or large algae-breeding and sustainable fisheries can be developed and new forms of energy supply can be realised where the energy richness is largest (e.g. wind turbines where wind blows, and a tidal plant where the tides rule.

8.5.4 Destructive Mob-Elections

The fourth scenario, destructive mob-elections (Fig. 8.23) has been developed in reaction on a very general question how to use Swarm Planning in regions. Main proposition is to make use of surprise and democratic flash-mobs in the planning of regions.

In this scenario the random destruction of buildings and/or villages as a result of surprise elections by text-messages is proposed. The idea is to organise elections by making use of text-messages within a certain area. Voters can vote on their favourite building to be demolished. This election can be organised instantly, on very short

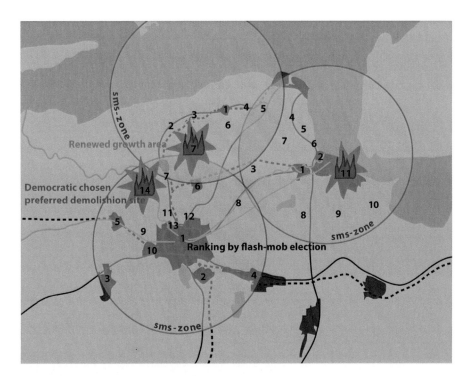

Fig. 8.23 Swarm model four: destructive mob-elections

notice and gives immediate results. The result, destruction of a certain building might sound harsh, but this randomised destruction also keeps a promise of new investments in the least popular places in a certain area. It emphasises also that instead of planning in detail for an entire region, interventions in certain nodes may lead to more development.

8.6 Conclusion

The examples of Swarm Planning design in this chapter illustrate that the Swarm Planning Theory and methodology can be used at, at least, four different levels.

The design can be directed by introducing a certain strategy. The advantage is that multiple interventions can be derived from one strategy, but a disadvantage is that the strategy can be interpreted in ways that do not emphasise a sustainable development or climate proofing.

The second level is the level of the intervention itself. The examples shown in this chapter demonstrate that a precise intervention can lead to widespread, and positive effects in the area around the intervention.

At the climate landscape level the approach proves to be most powerful. Not only the identification of (the location of) the intervention is important to anticipate future climate hazards, the design of the dynamic impact of the event as result of an intervention, support areas to not only prepare for the future but also direct the spatial changes in their environment. This strategy is functional in different types of climatic effects, e.g. both for flooding and sea level rise as for bushfires, as shown through the examples for Eemsdelta, Murrindindi and Bendigo.

The Swarm Planning Experiment, ultimately, shows the power of the approach when real policy boundaries are removed. In a laboratory setting, occupied by 30 intellectuals, unpredictable and unexpected results were reached. The triggering of brain cells through opening the spectrum of planning tools and challenges led to very innovative spatial solutions, sometimes even extreme ones. However, these examples illustrate the broad opportunities Swarm Planning offers.

References

2009 Victorian Bushfire Royal Commission (2009) Final report, summary. Parliament of Victoria, Melbourne
Deltacommissie (2008) Samen werken met water, een land dat leeft, bouwt aan zijn toekomst. Bevindingen van de Deltacommissie 2008. Deltacommissie, Den Haag
Hamilton C, Kaiser T (2009) Psychological adaptation to the threats and stresses of a four degree world. Four Degrees and Beyond Conference, Oxford
Hansen JE (2007) Scientific reticence and sea level rise. Environ Res Lett 2(April–June 2007):024002. doi:10.1088/1748-9326/2/22024002. The open access journal for environmental science. IOP electronic journals

Hansen J, Sato M, Kharecha P, Russel G, Lea DW, Siddal M (2007) Climate change and trace gases. Phil Trans Soc A 365:1925–1954. doi:10.1098/rsta.2007.2052

Hansen J, Sato M, Kharecha P, Beerling D, Berner R, Masson-Delmotte V, Pagani M, Raymo M, Royer DL, Zachos JC (2008) Target atmospheric CO_2: where should humanity aim? Open Atmos Sci J 2:217–231. doi:1874-2823/08

Lenton T, Held H, Kriegler E, Hall JW, Lucht W, Rahmstorf S, Schellnhuber HJ (2008) Tipping elements in the earth climate system. Proc Nat Acad Sci USA 105(6):1786–1793

Newman J, Al-Bazo S, Kendall W, Newton J (2011) Re-burn. Design Studio KINDLE, Landscape Architecture, School of Architecture and Design, RMIT University, Melbourne

Rahmstorf S, Cazaneva A, Church JA, Hansen JE, Keeling RF, Parker DE, Somerville RCJ (2007) Recent climate observations compared to projections. Science 316:709

Roggema R (2008) The use of spatial planning to increase the resilience for future turbulence in the spatial system of the Groningen region to deal with climate change. In: Proceedings UKSS-conference, Oxford

Roggema R (2009) Adaptation to climate change, does spatial planning help? Swarm planning does! In: Brebbia CA, Jovanovic N, Tiezzi E (eds) Management of natural resources, sustainable development and ecological hazards. WIT Press, Southampton, pp 161–172

Roggema R, Huyink W (2007) Atlas Groningen, analytical document environmental-spatial plan. Internal document. Province of Groningen, Groningen

Tin T (2008) Climate change: faster, stronger, sooner, an overview of the climate science published since the UN IPCC fourth assessment report. WWF European Policy Office, Brussels

TU Delft (2008) Het -1,3 meter plan. Hotspot Zuidplaspolder. Xplorelab, provincie Zuid-Holland, Den Haag

Chapter 9
Cities as Organisms

Andy van den Dobbelsteen, Greg Keeffe, Nico Tillie, and Rob Roggema

Contents

9.1	Background	196
9.2	The City as Organism	198
	9.2.1 Introduction	198
	9.2.2 A Definition of Life	199
	9.2.3 Features of Individual and Collective Life	199
9.3	Cities and Energy	202
	9.3.1 Cities Rather than Buildings	202
	9.3.2 Energy and Climate	202
	9.3.3 Excess and Shortage	202
9.4	Approaches to Becoming Autonomous	203
	9.4.1 The Rotterdam Energy Approach and Planning (REAP)	203
	9.4.2 Swarm Planning	205
9.5	Conclusion	205
References		206

This chapter has previously been published in the Proceedings of ICSU2010 (Teng 2010)

A. van den Dobbelsteen (✉)
Climate Design & Sustainability, Faculty of Architecture,
Delft University of Technology, Julianalaan 134, 2628 BL Delft, The Netherlands
e-mail: a.a.j.f.vandendobbelsteen@tudelft.nl

G. Keeffe
Sustainable Architecture, School of Planning Architecture
and Civil Engineering, Queens University Belfast, David Keir Building,
Stranmillis Road, Belfast BT9 5AG, UK

Leeds Metropolitan University, Leeds, LS6 3QS, UK
e-mail: g.keeffe@qub.ac.uk

N. Tillie
Faculty of Architecture, Delft University of Technology,
Julianalaan 134, 2628 BL, Delft, The Netherlands
e-mail: n.m.j.d.tillie@tudelft.nl

Abstract Since the UN report by the Brundtland Committee, sustainability in the built environment has mainly been seen from a technical focus on single buildings or products. With the energy efficiency approaching 100%, fossil resources depleting and a considerable part of the world still in need of better prosperity, the playing field of a technical focus has become very limited. It will most probably not lead to the sustainable development needed to avoid irreversible effects on climate, energy provision and, not least, society.

Cities are complex structures of independently functioning elements, all of which are nevertheless connected to different forms of infrastructure, which provide the necessary sources or solve the release of waste material. With the current ambitions regarding carbon- or energy-neutrality, retreating again to the scale of a building is likely to fail. Within an urban context a single building cannot become fully resource-independent, and need not, from our viewpoint. Cities should be considered as an organism that has the ability to intelligently exchange sources and waste flows. Especially in terms of energy, it can be made clear that the present situation in most cities are undesired: there is simultaneous demand for heat and cold, and in summer a lot of excess energy is lost, which needs to be produced again in winter. The solution for this is a system that intelligently exchanges and stores essential sources, e.g. energy, and that optimally utilises waste flows.

This new approach will be discussed and exemplified. The Rotterdam Energy Approach and Planning (REAP) will be illustrated as a means for urban planning, whereas Swarm Planning will be introduced as another nature-based principle for swift changes towards sustainability.

Keywords Sustainable development • Energy neutrality • Carbon neutrality • Biomimetrics • Organisms • REAP • Swarm Planning

9.1 Background

Although the earth receives almost 9,000 times more energy from the sun than that mankind needs, energy is becoming a huge problem. Western societies rely heavily on energy, fossil fuels in particular. The Netherlands for instance produces less than 4% of its energy by means of sustainable sources (CBS 2008). The rest is fossils and a bit of imported nuclear energy. As Mackay (2009) demonstrated, it is very difficult

---

R. Roggema
The Swinburne Institute for Social Research, Swinburne
University of Technology, PO Box 218, Hawthorn, VIC 3122, Australia

Delft University of Technology, Delft, The Netherlands

Wageningen University, Wageningen, The Netherlands

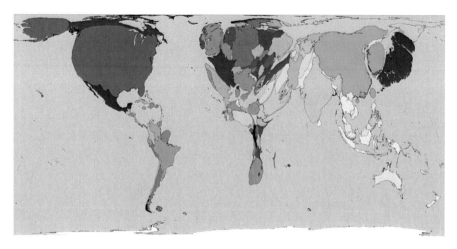

Fig. 9.1 Developed countries above the equator infest on other regions for energy... Countries and the area of land respective to the amount of fuel they consume (Dorling et al. 2009, downloadable from www.worldmapper.org)

to establish a society fully run on renewables. However, Cullen and Alwood (2010) showed that most of the energy we use is lost as non-functional waste energy. So the initial demand can be reduced by more effective usage, such as by low-exergy means (Stremke et al. 2011).

Although estimates of resources fluctuate, it is apparent to both energy experts and oil companies that the end is coming near. We have passed peak oil (ITPOES 2010): these days we consume more oil than can be produced. That this is a literally dangerous situation was demonstrated by the two gulf wars and recent turmoil around gas from Russia (first: Ukraine disconnected, second: Belarus threatening to halt the throughput of Russian gas). Apart from this international perspective and its influence on the price of energy, few people from the West understand how dependent they have become on energy, and that a collapse in the provision would have devastating effects to everyday life.

Last but certainly not least, the western hunger – or rather thirst – for energy is severely limiting the opportunities of developing and emerging regions to catch up in prosperity. As Fig. 9.1 indicates, western countries owe their prosperity to limited use of energy in other parts of the world. Needless to say this situation deviates strongly from the equity goals posed by the Brundtland Committee in 1987.

The abundance – until now – and relatively cheap and easy access to fossil energy has made the world lazy and inactive to search for local possibilities that would avoid demand from alien energy in the first place. We need to learn this again: planning and designing in such a way that local resources are optimally seized before any demand is posed upon other areas. Methods as Energy Potential Mapping (EPM) (Dobbelsteen et al. 2007) can support this.

In addition to and in relation with this new focus needed on local potentials, energy systems better based on (non-squandering) natural principles may support a shift towards more sustainable cities and regions.

9.2 The City as Organism

9.2.1 Introduction

Man has lived in cities for some 8,000 years. During this time cities have become "the most complex thing man has designed" (Richard Rogers and Kenneth Powell), as they draw in resources from many global sources and produce many streams of waste, at one level and function as highly tuned psycho-geographic entities at another. The urban dependence on lands outside the city borders has become irresponsible and will eventually lead to the city's demise, as many predecessors from the past have experienced (think of the Mayan cities and most of the original classical metropolises). With resources depleting, fossil fuels in particular, the need to become at least partly autonomous and less vulnerable is urgent.

In his book 'Creating Sustainable Cities', Herbert Girardet (1999) likened the modern city to a superorganism, rather like a bee-hive or termite mound, and it is this concept that this paper intends to develop. If we are looking for models of sustainability, then life having been on the planet for some 4 billion years should offer some models and ideas for our engagement with the ecology we find ourselves in. However 'life' itself is a complex notion and offers many examples of success from viruses to polar bears and trees. With so many different types of engagement by life-forms with their ecologies available, just how do we choose and use these varied exemplars to develop an ecological model of the function and form of the modern city.

Life can be seen, argues Keosian (1964) in his seminal text 'The Origin of Life', not only as an individual entity but also as a collective entity. Each view possesses different factors that define the key elements of living things. The superorganism is a particularly interesting concept as it possesses not only the factors of a single life-form but also those of life in a collective sense. This means that if the city is a true superorganism, then its structure and processes will be varied and complex, and nested at a range of scales.

The city may be complex and thus in compliance with the superorganism of Girardet, but it does not necessarily mean that it is an intelligent superorganism. The way in which most of our present-day needs are served resembles a system of infinite throughput without recycling or feeding loops to the places resources initially came from. And this is facilitated by a centralised system that hardly interacts with local circumstances, or reverse. In fact, the city functions as the 'intensive care of a hospital' (Dobbelsteen 2010): a collection of individual edifices, which fully rely on the central provision of water, food, materials, gas, electricity, telecommunication, sewage and other waste collection. Disposal of waste heat and discharge of rain water are not considered as the loss of valuable sources or as the burden shift to elsewhere, what these actually are.

Just as sustainable architecture – climatic design in particular – can learn from nature and become a servile form of biomimicry (Benyus 2002), cities and their systems of essential flows (Timmeren 2006) could learn from nature too and copy functional, practical and sustainable elements to become an intelligent anthropogenic superorganism.

9.2.2 A Definition of Life

Let us learn from natural life first.

The living autonomous city will contain both the factors of an individual life-form and those of life as a collective entity. Keosian (1964) defined individual life as an open system, powered by sunlight and having five key defining factors, namely Order, Energy, Separation, Self-perpetuation and Evolution, each being necessary in all living things. In addition to these, life is also a collective thing – working within small open system we call ecologies, this also has defining factors that are subtly different. These are: Order, Energy, Homeostasis, Cybernetic systems and finally Separation. Each of these factors needs to be present for a system to be considered 'alive'.

9.2.3 Features of Individual and Collective Life

9.2.3.1 Order

All living things exhibit order, compared to the chaos of non-living entities. The order expresses itself in the way of structure, and this structure manifests itself in the way of a synergy, where the sum is greater than the parts. In a collective sense, order is characterised by life's structure consisting of a complex system of interdependent organic systems, acting within a closed materials system.

Cities, from their very initiation, have been ordered, sometimes explicitly and sometimes by emergent forces. Good examples of the first are the grids of Manhattan or the *Eixample* of Barcelona. Less formally planned cities also exhibit order, such as Damascus, where courts and narrow streets create shade. This shade then provides cooling for buildings and also niches for various activities to take place. This exhibits the idea of synergy, where the juxtaposition of elements creates extra possibilities. This can be manifest in non-visual structures too such as exergy nets, or closed-cycle planning.

9.2.3.2 Energy

All living things expend energy as life is an open system directly dependent on the sun for its energy. Collective life can be seen as an unsustainable open system of energy, ultimately dependent on solar energy. This use of energy is manifest as a creature's metabolism, a cyclical process, which transports and transforms energy around the entity and makes it useful to do work. Secondly all living things store

Fig. 9.2 A bean plant does not squander its energy: it uses its finite energy source, contained in the bean, to build up a sustainable system of solar panels, the leaves

energy for times of hunger, as food is not always available, and this creates a rhythm based around solar cycles that promote activity and dormancy. In close interrelationship with the features mentioned, an intelligent energy system – which organisms that have survived over a long time demonstrate – needs to be responsive to alterations in climatic conditions, the weather and other circumstances. Homeostasis, comprising a myriad of mechanisms for instance to battle cold or heat, is a means to sustain itself under extreme conditions. Also for collective life it is the constant evolution to regulate the environment at a favourable condition.

In the modern city, the energy issues are more problematic in our definition of 'living' as fossil fuels – which are ancient sunlight - have led to very skewed practices. The living city will be a solar city, collecting, storing and transporting energy within it. This is entirely possible, considering that even in Manchester (United Kingdom, Latitude 54°N) enough energy falls on the city from the Sun in 1 day in June to power the city for a year. The issue is not the amount of energy available, but the methods of collection, storage and utilisation. Individual buildings need to be designed to collect the sun and not shade others (Keeffe and Martin 2007), and a light – stratified – system of development (like a forest) needs to be put in place. Moreover, intelligent cities will be responsive to alterations to circumstances that endanger the sustenance of its being and have mechanisms by which a stable situation can be pertained or returned to equilibrium (Fig. 9.2).

9.2.3.3 Separation

Life exists within and maintains defined boundaries, which act to mediate between the internal-external conditions. Every organism has a protective skin that separates the (vulnerable) inside from the outside. Apart from separation it is the direct intermediary between inside and outside, so interaction can be established best through the skin.

In cities, separation is not as apparent as with individual buildings: the building envelope plays an important role in the separation between indoors and outdoors and should be the building component where intelligent interaction can be performed between the two environments.

9.2.3.4 Self-Perpetuation

Without self-perpetuation living beings would not survive. It is the sole reason for existence: no organism lives to die; it lives to perpetuate its life and produce offspring. In that sense, sustainable development as defined by the World Commission on Environment and Development (Brundtland et al. 1987) has a direct natural, biological foundation.

Cities should be focused on self-perpetuation as well. Currently they are not. For the time being they still snooze under the comfortable condition of abundance outside the city, never questioning the limit to this. Only in the occasional event of misfortune (a power plant that temporarily hampers, a broken drinking water pipe) one experiences the malady of being powerless when dependent on centralised utilities. Self-perpetuation is the quintessence of the sustainable city.

9.2.3.5 Evolution

Standstill is backward development. Creatures need to develop, both as an individual and as a species. Growth not necessarily means more yet better, so evolution's main aim is to make an individual or species stronger. Since life is never constant, also cities need to evolve into better, more efficient, more resilient organisms.

9.2.3.6 Cybernetic Systems

In addition to these characteristics, collective life also has cybernetic systems, active-control systems utilising negative feedback systems, which maintain homeostasis. In a way the human system of a city region perhaps also has a similar cybernetic mechanism but only in a centralised, tardy way. Sustainable cities should have cybernetic control on various scales and responding rapidly to altering conditions (for instance the weather, supply of resources, over-production).

9.2.3.7 Symbiosis

In nature symbiosis occurs a lot. Organisms support one another, such as hippos cleansed from parasites by fish, which in their turn live of the nutrient-rich manure from hippos. This perhaps may not seem easily translatable to cities, but this kind of interaction could be superposed on the use of energy. In cities a lot of energy is used and wasted without symbiosis. The following will describe an urban energy system that is more in accordance with intelligent organisms.

9.3 Cities and Energy

9.3.1 *Cities Rather than Buildings*

Urban design and the fundamental principles of how to shape our cities, has so far barely featured in the greenhouse debate. The urban dimension and the macro-scale of cities were mostly missing in the debate of the 1980s and early 1990s, as sustainability was predominantly discussed as being about 'alternative lifestyles'. Much of the more recent debate has circled around ideas about active technology for 'eco-buildings' and sophisticated façade technology - rather than about urban issues. Nevertheless, "sustainable architecture is only really effective when set in an urban planning context which itself is based on sustainable principles" (Gauzin-Müller and Favet 2008).

9.3.2 *Energy and Climate*

The energy system in our cities is mostly based on our winter climate and appears not to be flexible when it comes to seasonal changes. Climate change will bring a shift from mainly heat demand to more cooling demand, since winters will become milder and hot summer days will increase. To be able to maintain comfortable indoor climates the energy system in a city should also be designed on the summer cooling demand.

9.3.3 *Excess and Shortage*

The way cities consume energy can be very well described as autistic. Only high quality energy is used as input, while many processes can do with lower quality energy. This results in high energy-consumption with a high amount of waste energy. In an intelligent system this waste heat can directly be input for another function. If there is a surplus of heat in the summer season heat can be stored in aquifers and used again in winter.

9 Cities as Organisms

Fig. 9.3 Energy demand by different urban functions (W = heat, K = cold, E = electricity): different patterns occur at the mean time, causing unnecessary use of energy (Image by DSA)

Another way to increase energy efficiency from sustainable recourses is to respond more to the supply rather than the demand. During windy or sunny periods more energy is available, but because the net cannot deliver extra energy and the cost to store this are too high, wind turbines are halted at a certain speed. Chargeable batteries connected to the net can have an intelligent interface, which starts charging when there is a surplus of energy. An extra stimulation can be given to vary the price according to the available supply. In this way laptops, phones, electrical bikes and in the coming years also cars can function together as a large battery for wind energy (Fig. 9.3).

9.4 Approaches to Becoming Autonomous

9.4.1 The Rotterdam Energy Approach and Planning (REAP)

The New Stepped Strategy (Dobbelsteen 2008), based on the three steps strategy or Trias Energetica (Lysen 1996), runs as follows:

1. Reduce consumption (using intelligent and bioclimate design)
2. Reuse waste energy streams
3. Use renewable energy sources and ensure that waste is reused as food
4. Supply the remaining demand cleanly and efficiently

Of which the latter step may be forgotten when design for a future that will lack these finite resources (Fig. 9.4).

This way it advocates optimal use of waste streams not only for each individual building but also on a citywide scale. Waste streams from one chain may be used in a different chain. For example, wastewater can be purified and the silt fermented to form bio-gas which can be reused in the energy chain.

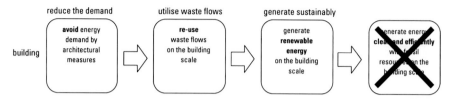

Fig. 9.4 The new stepped strategy

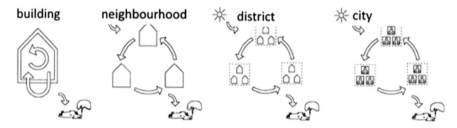

Fig. 9.5 Reuse of waste flows and upscaling from building to city level

9.4.1.1 Exchanging and Cascading Energy

According to the 1st Law of Thermodynamics, energy is never lost, but the second Law describes the increase of entropy, implying the decrease of something else. This is exergy. Exergy is a combination property of a system and its environment because unlike energy it depends on the state of both the system and environment.

When we look at the way our energy system is organized, then you see that 98% of the energy that is consumed is primary produced energy from fossil and nuclear sources. While at the final stage a lot of waste heat is lost in the air, water and soil.

High quality energy like natural gas that has a temperature of 1,200–1,500° should be used for processes that actually require this temperature like some industrial processes. But for the heating of a house to 20° this is not necessary. If the energy system of a house is well designed a temperature between 25° and 40° is sufficient. Therefore waste heat from greenhouses or from supermarkets can heat dwellings and functions, which need higher quality energy can be supplied by even higher processes. In such a low-ex system the consumption of primary energy is much lower because only the highest quality functions use fossil and nuclear energy. A system of cascading of energy qualities can improve our energy system by a factor of 6 (Tillie et al. 2009).

These principles of reusing waste flows from one function in others, which can still profit from them and can be extended by exchanging within neighbourhoods, districts and the entire city, serving the symbiosis principle of nature (Fig. 9.5). This is the key message behind the Rotterdam Energy Approach and Planning (REAP), which is discussed more elaborately else in these proceedings (Tillie et al. 2010), which is used in the Dutch *mainport* city of Rotterdam, yet also in other cities that follow the example. The REAP method is currently even converted to a similar method for other flows than energy: water, materials and food.

9.4.2 Swarm Planning

Not just our consumption of energy demands for radical changes. Climate change is affecting our natural, social and technical systems fundamentally. These changes take place over longer periods and emerge slowly. In contrast, current planning systems focus on short periods and try to enforce immediate changes.

Meanwhile, society is becoming more complex, as are developed ecosystems in nature themselves. This complexity of interactions requires a spatial system to change its conditions rapidly and frequently. In contemporary planning practice designs are mostly fixed images of the future.

To realise energy-autonomous and climate-proof designs in an increasing complex society a new planning paradigm must be developed, which is capable of integrating long term aims and is flexible enough to respond to the complex society. This regional or urban planning approach needs to incorporate future developments and let these single developments influence the entire regional system. Main objective is to make use of these individual changes, instead of trying to reduce the effects of changes. The new planning methodology can be called Swarm Planning (Roggema and Dobbelsteen 2008).

Swarm Planning is a metaphor for a planning method in which the aim is to give incentives at certain points, which change the entire region and society, just like a swarm of birds that changes its shape suddenly as a result of one simple impulse. In this sense Swarm Planning is another unlikely biomimetic principle to be applied to planning.

The hypothesis is that planning in a swarm-way is very useful in complex systems and if the issues are long term oriented. Insights from both complexity theory and marketing are used to build up this new planning method. For Swarm Planning significant parameters of climate change and energy provision need to be thoroughly analysed by their spatial impact. Hence, mapping them can be helpful. The major twist is made in defining the right impulses to adapt the urban or regional society to foreseen changes. Network analysis is one of the tools to be used then.

Swarm planning is currently elaborated and tested on various locations and situations.

9.5 Conclusion

We are only at the start of comprehending natural principles related to life and translating these to the techno-sphere we create: products, buildings and cities. Especially cities can learn from life in nature and copy characteristics that make organisms survive and evolve. This is particularly so with the current and upcoming developments in regards to the availability or resources, such as fossil fuels.

A sustainable city can and should transform from a collection of individual buildings, which jointly rely fully on the provision of essential flows from a distant centralised facility, to an intelligent organism that has all features of life of individual and collective organisms. In this paper we attempted a first step to translate theory on this to urban principles and exemplified it through the Rotterdam Energy Approach & Planning (REAP) and Swarm Planning. It is but a modest start, however important considering the immense changes needed for cities to become sustainable.

References

Benyus JM (2002) Biomimicry – innovation inspired by nature. Harper Perennial, New York
Brundtland GH, World Commission on Environment and Development et al (eds) (1987) Our common future. Oxford University Press, Oxford/New York
CBS (2008) Duurzame energie in Nederland. CBS, Heerlen
Cullen JM, Allwood JM (2010) The efficient use of energy: tracing the global flow of energy from fuel to service. Energy Policy 38(1):75–81
Dorling D, Newman M, Barford A (2009) The atlas of the real world – mapping the way we live. Thomas & Hudson, New York
Gauzin-Müller D, Favet N (2008) Sustainable architecture and urbanism: concepts, technologies, examples. Birkhauser, Boston
Girardet H (1999) Creating sustainable cities. Schumacher briefings no. 2, Green Books Ltd. Devon, UK
ITPOES (Industry Taskforce on Peak Oil & Energy Security) (2010) The oil crunch – a wake-up call for the UK economy. http://peakoiltaskforce.net. Accessed Feb 2010
Keeffe G, Martin C (2007) Bioclimatic architecture labs. In: MSA yearbook. Manchester School of Architecture, Manchester
Keosian J (1964) The origin of life. Chapman & Hall, New York
Lysen EH (1996) The Trias Energetica – solar energy strategies for developing countries. In: Proceedings of the Eurosun conference, Freiburg
MacKay DJC (2009) Sustainable energy – without the hot air. UIT Cambridge Ltd., Cambridge
Roggema R, van den Dobbelsteen A (2008) Swarm planning – development of a new planning paradigm, which improves the capacity of regional spatial systems to adapt to climate change. In: Proceedings of the world sustainable building conference (SB08). CIB/CSIRO, Melbourne
Stremke S, van den Dobbelsteen A, Koh J (2011) Exergy landscapes: exploration of second-law thinking towards sustainable spatial planning and landscape design. Int J Exergy 8(2):148–174
Teng (2010) Proceedings of the First International Conference on International Conference on Sustainable Urbanization (ICSU 2010). Faculty of Construction and Land Use, The Hong Kong Polytechnic University, Hong Kong, China, 15–17 Dec 2010, 290 pagina's
Tillie N, van den Dobbelsteen A, Doepel D, de Jager W, Joubert M, Mayenburg D (2009) Towards CO2 neutral urban planning – introducing the Rotterdam Energy Approach and Planning (REAP). J Green Build 4(3):103–112
Tillie NMJD, Doepel DR, van den Dobbelsteen AAJF (2010) Towards carbon neutral city planning: Rotterdam Energy Approach and Planning (REAP) – a tool for urban planners in accommodating flows of energy, water, material and natural resources. In: Proceedings ICSU2010. Hong Kong Polytechnic, Hong Kong
Van den Dobbelsteen A (2008) Towards closed cycles – new strategy steps inspired by the Cradle to Cradle approach. In: Proceedings PLEA2008, Dublin
Van den Dobbelsteen A (2010) Use your potential! – sustainability through local opportunities. TU Delft, Delft
Van den Dobbelsteen A, Jansen S, van Timmeren A, Roggema R (2007) Energy potential mapping – a systematic approach to sustainable regional planning based on climate change, local potentials and exergy. In: Proceedings of the CIB world building congress 2007. CIB/CSIR, Cape Town
Van Timmeren A (2006) Autonomie & heteronomie. Eburon, Delft

Chapter 10
The Best City?

Rob Roggema

Contents

10.1	Introduction	208
10.2	Best *Planned* Cities	216
	10.2.1 Chandigarh, India	216
	10.2.2 Brasilia, Brazil	219
	10.2.3 Almere, The Netherlands	222
	10.2.4 Canberra, Australia	226
	10.2.5 Communalities	228
10.3	Best *Sustainable* Cities	229
	10.3.1 Freiburg, Germany	229
	10.3.2 Curitiba, Brazil	233
	10.3.3 Malmö, Sweden	236
	10.3.4 Masdar City, United Arab Emirates	238
10.4	The *Self-Organising* City	241
	10.4.1 Delhi, India	244
	10.4.2 Cairo, Egypt	244
10.5	The *Best* City, Delhi Designs for Hobsons Bay	246
	10.5.1 Gagan City	247
	10.5.2 GEO City	247
	10.5.3 Cloud 9	249
10.6	The Art of Designing for Climate Adaptation	253
References		256
Websites		256

Abstract Designers have long tried to design the best possible city. Many examples illustrate however that buildings and cities are although designed with the best purposes, in reality not function very well. The local climate is often difficult to

R. Roggema (✉)
The Swinburne Institute for Social Research, Swinburne
University of Technology, PO Box 218, Hawthorn, VIC 3122, Australia
e-mail: rob@cittaideale.eu

influence the designs in a way that these buildings are successful, even more needed in a changing climate. Explorations of the past planned, sustainable and self-organising cities illustrate that despite the fact that these cities were developed with honest and valuable goals, there are many side effects that contradict with the original aims. When smart people from another culture, without constraints about regulations and political habits are asked to design the best city, a wide range of climate adaptive strategies are implemented in the designs.

It is clear that the best city doesn't exist, but six actions are distinguished that enhance climate adaptive cities: analyse networks, focus on key nodes, free self-organising developments, plan the unplanned, release control and check-up regularly.

Keywords Best city • Sustainability • Art of designing • Self-organising • Climate adaptation

10.1 Introduction

Recently I visited Delhi. Dr. Sundaresan Pillai of CSIR invited me to present the Diamond Jubilee Lecture. During my stay, I had a very pleasant dinner with my friend Sanjay. We talked about many subjects, amongst which how to increase sustainability as designers. Sanjay told me about one of his projects, in which the design for new schools was undertaken in full collaboration with the future users. We concluded that thorough sustainability could only be achieved when people are given the opportunity to internalise sustainability goals, measures and strategies. In our opinion, this could only be achieved if designers support the future users of the design to envisage, design and build their own desirable future. A designers' role as the person who is delivering his personal piece of architecture from *heaven*, bringing the best design according to *his* own standards to the people that are supposed to use its creation, is often contra-productive. Only time-consuming investments in local people will internalise aims of sustainability and hence deliver designs that are developed locally and reflect the locally and individually felt best and most sustainable solutions. These designs are filled with local knowledge about the environment, culture and local desires, knowledge that cannot be grasped by any designer from the outside. However, to be able to support local groups to invent their own sustainable design solutions is intensive and very time consuming albeit rewarding too. The way we were identifying a best practice in sustainable design resonates with the design charrette approach (see Chap. 7), which requires the designer to position himself out of the limelight, substantial time investments and building of relationships and the ability to work while there is no certainty beforehand about the outcome of the design process. It is often easier to hire a famous architect to deliver his piece of artwork .... When taking a look at several recent examples of design the question can be asked how specific and locally embedded current architecture and city planning and design really is?

10 The Best City?

Fig. 10.1 Two pieces of architecture: Lotus Temple and Opera House. Or: Opera House and Lotus Temple? (Photo's: Rob Roggema)

The two pictures in Fig. 10.1 were made within 1 week. Still, it is not the same building. One of the pictures shows a part of the Lotus Temple in Delhi and the other does the same from Sydney's Opera House, both beautiful pieces of architecture. When looking at the whole building (Fig. 10.2) it becomes clear which building is located in India and which one in Australia. These buildings could be brother and sister twins, because they look very similar. The peculiar part of this is that the similarities in both buildings are not underpinned by local climates nor can it be culturally explained. The hot and sticky summer in the Delhi metropolis cannot be compared with Sydney's summer sea breeze, which, even on the hottest days of the

Fig. 10.2 The Lotus Temple and the Opera House (Photo's: Rob Roggema)

year, has a pleasant cooling effect. The shape of the buildings and the choice for their white skin has most likely not been influenced through the local climate.

This example does not stand alone. The two images of Chinese pavilions in Fig. 10.3 illustrate that huge similarities are realised apart from the climatic or local context. Both the hotel in the Dutch village of Breukelen as the pavilion in Melbourne's inner western suburbs look alike, but have to deal with fundamentally different climates.

Even more extravagant are the two hotels in the Amsterdam metropolitan area (Fig. 10.4). The *Sea palace*, again, exhibits the properties of a Chinese pavilion, and is adapted to its environment (the Amsterdam harbour) by making the building

10 The Best City?

Fig. 10.3 Breukelen The Van der Valk Hotel (http://www.rederijvonk.nl/salonboot-varen-vecht-omgeving/hotels-aan-de-vecht.html) and the Chinese pavilion in Melbourne's Inner Western suburb (Photo: Rob Roggema)

floating. The other hotel, Inntell in the Northern suburb of Zaandam is a high-density aggregation of, artificial, old houses and facades that used to be built in the area. In a certain sense both hotels have been adapted to the local environment or history. The question however, may be raised how locally adapted these buildings really are, or, did local and future users of the building and urban environment have any

Fig. 10.4 Sea palace Amsterdam (http://www.worldisround.com/articles/331017/photo28.html) and Inntel hotel Zaandam/Amsterdam (http://www.happy-pixels.com/2011/06/08/zaandam-hotel/)

influence on the design, or have they been offered a role as co-designers of the project? The most likely answer in both these cases is negative.

When Le Corbusier, who was born as Charles-Édouard Jeanneret, designed the buildings for the Capital Complex in Chandigarh, he designed them as if they were to be built in Paris. The enormous thermal mass lets the buildings function in the hot Indian summer as stoves. It makes air-conditioning a necessity. Many facades are therefore, against the original intentions of Le Corbusiers design, covered with air-conditioners, such as on the Assembly building (Fig. 10.5).

The English style Victorian housing dropped in Melbourne has difficulties dealing with the Australian hot summers, as the English designed and build them after

10   The Best City?

Fig. 10.5 The Assembly building in Chandigarh (Photo: Rob Roggema)

Fig. 10.6 Victorian housing in Melbourne's inner west (Photo: Rob Roggema)

examples from home in the early days, and they were optimal prepared to deal with rain (Fig. 10.6).

Or take a more recent example, the design for the Etihad stadium in Melbourne's Docklands (Fig. 10.7). This huge stadium is oriented as if the sun shines from the south: the European architect assumed (or just forgot) that in the Southern

Fig. 10.7 Etihad stadium, Melbourne (Photo: Rob Roggema)

Hemisphere the sun isn't similarly oriented as in the Northern Hemisphere. As a consequence, the grass in the stadium has difficulties to grow.

These random examples illustrate that, in several occasions, in current times and in history and by famous architects, the local climate is not always taken as the basis for the design, leading, in many occasions to serious suboptimal conditions. Other considerations were probably more important, such as habit, the drive to showcase new architecture, or just forgotten. When taken beyond architecture, the question may be raised what the best city is, if at all this can be defined.

This question is not new. In 1516 Thomas More described his island Utopia (Fig. 10.8) (Logan and Adams 1989) and many followed. In the book 'Ideal Cities: Utopianism and the (Un)Built Environment', Eaton (2003) describes them all. Rem Koolhaas' with his 'The voluntary prisoners of Architecture', from Pieter Breugels painting 'Luilekkerland' to Archigrams 'Instant City' and Superstudio's 'Continuous Monument' and everything in between: Medieval miniatures of heavenly cities, geometrical ideal- and fortified cities dating back to the Renaissance, American grid-cities, the Chaux by Ledoux, The Phalanstere of Fourier, Port Sunlight, Ebenezer Howard diagrams, Paul Citroens 'Metropolis collage', Expressionistic, Constructivist and Futuristic cities from the Twenties, Hugh Ferris' chalk drawings of an idealised New York, Le Corbusiers plans for Parijs, Frank Lloyd Wright Broadacre City, New Babylon or Friedmans' Space Frame Cities.

Fig. 10.8 The island Utopia by Thomas More (1516)

But as many tried to design and describe the ideal city, even so many failed. How can we describe the ideal city? This question, one of the most difficult to answer, is subject of this, final, chapter. Some clues can be identified. The ideal city must be sustainable, but not dogmatic. It must be clean and safe, but not boring. And the ideal city must strive for order, but in a flexible way. One element unifies the former. Rules are not put in place to determine the design interventions, but it's the other way around. The intervention is the rule. The intervention rules and directs developments. According to the Scientific American, this future, better, greener and smarter city is urban (DiChristina et al. 2011). The role of social (cyber) networks in increasing connectivity and faster solutions is emphasised (Ratti and Townsend 2011), as well as the potential of shantytowns, slums and favela's as the places where creativity and innovative solutions find their origin (Neuwirth 2011). The key message is that new and sustainable solutions can be expected from these unexpected places more than from the established and well-organised western world. Again, where rules do not hinder creativity and the intervention is the rule.

In this, final, chapter explorations on the 'best city' lead us to cities that were planned from scratch, the ideal starting point to design the best possible city, to cities that were planned with the specific aim to create a sustainable city, and to cities that emerged, and were not planned. Secondly, the findings of a design session aiming to design the best climate adaptive city are discussed. The chapter concludes with a first description of the art of designing for climate adaptation.

10.2 Best *Planned* Cities

10.2.1 *Chandigarh, India*

The city of Chandigarh in the north of India was developed as the Capital of India in the 1950s of the twentieth century. The design of the Masterplan (Fig. 10.9) and several of the major buildings in the city were designed by Le Corbusier.

The city is subdivided in sectors of the same size. The clear pattern is very readable, buts especially from the map or drawing board. Because the topography is not very accentuated, the repetitive pattern doesn't work as the regular system,

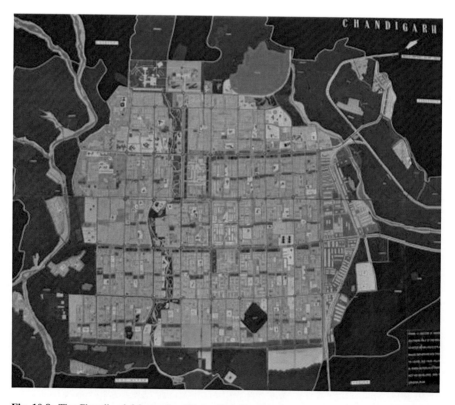

Fig. 10.9 The Chandigarh Masterplan (Photo: Rob Roggema)

10 The Best City?

Fig. 10.10 Typical street in Chandigarh (Photo: Rob Roggema)

within which the topography causes the exceptions. The street pattern consists of broad streets with a surplus of green and tress alongside them (Fig. 10.10).

Within the rigid urban pattern a couple of exceptions are contributing to the readability of the city. The riverbed to the West as well as the Capital Complex area in the North both function as points for orientation. The city centre (sector 17) is well known by everyone in the city, but can't be recognised as the centre of a large city. It appears as just another sector and intensified use or high-rise buildings one would expect in a city centre aren't there. The main quality of the city centre is the large open public spaces (Fig. 10.11) and the fact that most of the shops are found here. Furthermore, ground level is split in order to separate pedestrians and car traffic (Fig. 10.11). Because of the sizes of the superblocks, distances in the city are large and a car is required for transportation. This results in large spaces for car parking around the city centre.

The Capital Complex is designed as a free open space with big buildings amidst it. Several of Le Corbusiers most famous buildings are realised here, such as the Secretariat, the Assembly and the High Court (the latter two are in Fig. 10.12). Because of the fact that Punjab, of which province Chandigarh is the current capital, has to deal with political tensions, the free open space is hardly free. Many security points limit free wandering.

The city of Chandigarh cannot be compared with any other Indian city. It is a lush and very spacious city, where large public spaces, roads and green dominate the city lay out. Due to the distances the city is hardly walkable and ordinary Indian forms of transportation take over the road: auto-tuktuk, rikshas and cars. The accompanying horning reminds one of being in India. The iconic buildings are big and seem to be designed to impress. After more than 50 years of existence, Chandigarh

Fig. 10.11 Images of Chandigarh's city centre (multi-level crossing (*top*) and public space (*bottom*)) (Photos: Rob Roggema)

is ready for a makeover. In order to showcase its original values it needs to refurbish and refresh its public spaces, reconstruct and redesign many of its buildings and it needs rethink its traffic strategy: it could replace current road use (partly) by clean, fast and cheap public transport.

The climate in Chandigarh, especially in summer is hot and sticky. The city design, with its wide boulevards draped with trees provides shade and cooling. Water is not abundantly apparent in the city. However, the cities spacious layout gives Chandigarh a better liveability than most Indian cities.

10 The Best City?

Fig. 10.12 Iconic buildings in Chandigarh, designed by Le Corbusier: Assembly building and High Court (Photo's: Rob Roggema)

10.2.2 Brasilia, Brazil

The city of Brasilia was founded in the late 1950s of the Twentieth century and became Brazil's capital in 1960. The city was planned from scratch and gave designers, amongst them the lead designer Oscar Niemeyer, a free hand of designing the best city possible. The Masterplan followed all the rules of modernism: large blocks, lots of green spaces and separated traffic types. The grand design consisted of an urban pattern, which was shaped as a bird (Fig. 10.13).

Many of the residential neighbourhoods (Fig. 10.14) were built up around the traffic networks of wide roads, separated by green, all placed at respectable distances from city centre, amenities and each other. A car is almost a basic condition to live here. The residential blocks were lifted from the ground in order to let public green

Fig. 10.13 Night image of Brasilia from space (http://earthobservatory.nasa.gov/IOTD/view.php?id=48963)

Fig. 10.14 Typical neighbourhood in Brasilia (http://meredithinbrasilia.blogspot.com/2011/06/images-of-our-soon-to-be-neighborhood.html)

10 The Best City?

Fig. 10.15 The huge central axis in Brasilia (Photo: Rob Roggema)

'flow' underneath the buildings. This, supposedly, was designed to give the city an open and transparent character.

The most spectacular part of the city is undoubtedly the central axis (Fig. 10.15), which forms the body of the bird and alongside which all government buildings are found. The enormous dimensions of this axis do let the building look like regular sized ones, but in reality each of these buildings are multi-storey buildings of respectable sizes. The length of the central part of the axis alone is approximately 3 km, a distance not easily walked, especially not while crossing busy traffic on wide roads and wandering through a completely open space.

Brasilia houses many iconic buildings (Fig. 10.16). Most of them are located next to of near the main axis. The buildings are positioned as single buildings in wide-open spaces, allowing the building to make an impression. Because of the large size of the buildings they are placed much further away than expected. Oscar Niemeyer designed many of these buildings.

Brasilia is as a city to live in not very popular amongst government officials. Everyone who could afford it leaves the city during the weekends for better places, along Brazil's coast. This leaves many of the poor behind in the city, leading to Brasilia's favelas developing at a distance (approx 30 km) from the city. The large public spaces are difficult to maintain and many of these areas are no-man's-land. The huge distances require a car. Public transport or bicycle use is underexploited and the city could increase its densities through infill projects in the vast, unused open spaces.

Brasilia's climate is hot and humid, especially in summer. This imposes a high rate of air-con use. There are not too many trees, so shade can only be found under the buildings, where anonymity and criminality reign.

Fig. 10.16 Iconic buildings in Brasilia: Supremo Tribunal Federal and Memorial JK (Juscelino Kubitschek), both designed by Oscar Niemeyer (Photos: Rob Roggema)

10.2.3 Almere, The Netherlands

The city of Almere, in the Netherlands is built, started in the mid-1970s, in one of the Dutch polders, land reclaimed from the sea. Almere was designed as a polynuclear city, consisting of several urban districts (Almere-Stad, Almere-Haven, Almere-Buiten, Almere-Hout, Almere-Poort) which are separated from each other by robust green structures (Fig. 10.17). This causes an urban pattern of widespread developments. For instance, the distance between the northernmost and southernmost freeway exit is 13 km, for a city of nearly 200,000 inhabitants a long way.

Fig. 10.17 Almere city from the air (http://farm5.staticflickr.com/4126/4967385433_e97c1f2a0b_z.jpg)

Because distances between urban parts are large, car dependency is huge. At a close distance from Amsterdam it serves as the major residential area for the capital. Because many Almerians still work in Amsterdam, this causes a major traffic problem, especially because everyone needs to cross in mornings and afternoons a single bridge connecting the polder with the mainland.

The layout of Almere was originally based on two interfering systems: the urban design principles and the traffic network (Stassen 2001). The fine mazes of the urban design, with repetitive turns in the urban pattern of 45°, allowing for a fluent, unconscious flow from outside neighbourhoods towards the city centre. The traffic system was designed in a hierarchical way, starting from the freeways up to the individual streets. Super-positioned over this traffic system, the free-lying bus-lanes are realised, providing safe and fast public transport around the city. Later, the urban design became more linear (especially in Almere-Buiten and later phases of Almere-Stad), reflecting the underlying rationality of the polder. In the design for Almere, there is no 'central axis' or other spatial mega structure included. However, the architectural richness both elaborated in iconic buildings, such as for instance in the renewed city centre (Fig. 10.18) as in the design for the many individual homes, is unequalled in the country.

The vast majority of the houses in Almere are attached low-rise buildings, with entrance at ground level. The streets are neatly designed as coherent ensembles with architectural attention to the buildings and streetscape (Fig. 10.19). The houses all have a small garden in the front and a larger garden at the back. The pace of urban development in Almere is respectable. In many years of the past two decades the

Fig. 10.18 Iconic buildings in Almere, The Wave, by René van Zuuk Architects (http://www.e-architect.co.uk/holland/block_16_almere.htm) and Popzaal 2004 by Wil Alsop (http://straatkaart.nl/1315SB-Schipperplein/media_fotos/almere-centrum-Eyz/)

'production' of newly built houses was around 3,000 houses per annum. It has led to a population, which will soon top 200,000 people and it is expected to grow further to 350,000 inhabitants (Gemeente Almere 2009).

The Dutch climate is not very extreme and predictions will not lead to extreme disasters. Temperature will stay moderate (average highs 20–25 in summer, and zero–5° in winter) and despite the fact that relatively more intense rainfall is expected in summer, this will not lead to huge flood disasters. The main real threat to the

Fig. 10.19 Typical streetscape in Almere: Almere Pampus (http://www.flickr.com/photos/honorata_grzesikowska/4586421398/) and Almere Buiten (http://straatkaart.nl/1339GP-Chamoisstraat/media_fotos/chamoisstraat-almere-buiten-1cY/)

country is sea level rise as much of the country is lying below sea level. Almere, built in a deep polder, is around 5 m below current sea level. It borders an artificial lake (the IJssel Lake), which is separated from the sea through a closure dam. This lowers the risk of flooding as result of storm surges, but should the dam fail, the city is at risk.

10.2.4 Canberra, Australia

The city of Canberra is Australia's capital since 1927. In 1908 the site was selected, diplomatically situated between Melbourne and Sydney. Walter Burley Griffin designed the layout as the winner of an international design competition. The designs main elements are considered the large artificial lake and the main axis (Fig. 10.20), which visually connects Parliament House and Anzac Parade. The city is widespread and distances are large, leading to car dependency, similar to the other planned cities. Its natural features of mountains and forests are surrounding the city neighbourhoods.

Canberra hosts several iconic buildings, of which Parliament house and the National Museum of Australia are the most famous ones (Fig. 10.21).

Fig. 10.20 Main axis in the design for Canberra (http://www.agd.com.au/directory.php?dirpage=search&act=search&region_id=&cat=000084&keywords=&alpha_search=N)

10 The Best City?

Fig. 10.21 Iconic buildings in Canberra, National Museum of Australia (design: Howard Raggatt) and Parliament House (Mitchell/Giurgola Architects) (Photos: Rob Roggema)

The majority of neighbourhoods consist of suburban 'sprawl'. Detached houses are placed in lush and spacious patterns, with broad streets, green spaces and lots of trees (Fig. 10.22).

The climate of Canberra in winter is, due to its location in a mountainous environment, relatively cold and it can easily freeze. In summer however, temperatures rise and it can become very hot. In periods of longer droughts, combined with the abundantly available green and forests, Canberra is vulnerable for bushfires, as the most recent occasion in 2003 has shown.

Fig. 10.22 Typical streetscape in Canberra (http://treelogic.com.au/facts/2010/09/ageing-tree-avenue-management/, http://www.hotkey.net.au/~krool/photos/act/canberra.htm)

10.2.5 *Communalities*

The four cities that were planned from scratch share some communalities.

The first observation is that insiders are all very satisfied with the living conditions in their city. This information, only from hear-say, has been obtained listening to stories people have told about their cities. Except for Brasilia, people are happy and do not want to switch to another residence.

Secondly, the cities are in their design over-dimensioned. Sometimes in the form of (too) broad roads, enormous public spaces or long distances to travel from one end of the city to another. In several examples the idealistic enthusiasm to design the well-planned city carried away the designers. Especially the bigger buildings, requiring space around them to 'fit', lead to a scale in which human standards are in a sense set aside in favour of the overall picture. As humans you are little compared to the buildings around you, and to walk the city distances are always longer than you think.

As a result of this, and perhaps because these cities were designed in an era oil became abundantly available and the car was the mean to empower and mobilise large parts of the population, the four examples are all very car-dependant. The designs take the car and its requirements (road widths, parking space, long distances) as the prioritised element.

Thirdly, the Masterplans all became icons of good city planning. Due to their conceptual strengths, being central axes, grid patterns or poly-nuclearity, and being designed by some of the most famous architects and city planners, these cities are known all over the world. The emphasis on city planning subsequently, or in some cases parallel, led to increased attention for architecture. In all four examples innovative and iconic new buildings are designed and realised.

10.3 Best *Sustainable* Cities

The sustainable city, as can be expected, tries to create cities that are sustainable. Sustainability used to be a single goal issue, namely to create an environmentally better situation. This early awareness of sustainability was necessary because pollution was a threat to the lives of many people. Cleaning the soil, water, minimising air pollution or traffic noise were high priorities in those days. In the modern time the approach is holistic and many aspects of city development are integrated in a sustainable approach. Five components of a sustainable city are distinguished (Kassenaar 1994): ecological, socio-cultural, economic, spatial and infrastructural, and cybernetic components need to be sustainably integrated in the design of our future cities. Long standing examples, such as Curitiba and Freiburg as well as recent examples, such as Malmö and Masdar all pursue this goal of integration.

10.3.1 Freiburg, Germany

Freiburg was rebuilt after WWII and the city has chosen not to put the automobile central, but the pedestrians. This has led to streets deliberately designed for pedestrians, and bike lanes and trams (figure 10.23, at the heart of the city's development). Additionally, the unique Freiburg *Bächle*, small canals run down each central street.

Fig. 10.23 Freiburg City centre, crucial role for public transport and pedestrians (http://extension.ucdavis.edu/unit/environmental_management/course/description/?type=I&unit=ENV&prgList=GBD&course_title=Sustainability%20Abroad:%20What%20Can%20We%20Learn%20from%20Europe%20and%20the%20Rest%20of%20the%20World?&CourseID=36562)

As a result of this choice of half a century ago, 70% of the locals walk, cycle or take public transport (Fig. 10.23).

10.3.1.1 Traffic

Freiburg's traffic and transport policy gives preference to environment-friendly modes of movement (pedestrian traffic, cycling and local public transport). The most important objective of Freiburg's policy is traffic avoidance. This is achieved through the design of a compact city, which can be crossed quickly with. Urban development is directed along main public transport arteries and priority is given to centralised development over peripheral growth. All major urban developments follow this concept. As an example, the new city districts of Rieselfeld and Vauban are both easily accessible by public transport, as are the inner-city university locations.

10.3.1.2 Solar City

With more than 1,800 h of sunshine each year and an annual radiation intensity of 1,117 kilowatts (kW) per square-meter, Freiburg is one of the sunniest cities in

10 The Best City?

Germany. Not surprisingly, Freiburg has been most successful in the field of renewable energy. Solar panels can be found on the roofs of the Badenova Stadium and the City Hall, on schools, churches and private houses, on facades and towers. It has led to unique projects, such as the world's first energy self-sustaining solar building, the Heliotrope, the solar village created by Rolf Disch, or the zero-energy houses of the Vauban neighbourhood. Even the local football stadium has become an attraction as the first stadium worldwide to have its own solar plant (Fig. 10.24).

These principles of traffic and solar are implemented in current urban developments, such as Vauban and Rieselfeld.

10.3.1.3 Vauban

The Vauban Quarter is located close to the city centre, and is an attractive, family-friendly neighbourhood, in which civic commitment, collective building, and living with ecological awareness has great importance. The following eco-characteristics are realised:

Fig. 10.24 Examples of the use of solar energy in architecture and building: The soccer stadium (www.solar-fabrik.de/fileadmin/user_upload/pressebilder/referenzanlagen/SC_Stadion.jpg), Solar fabric (http://www.worldchanging.com/archives/011173.html) and Vauban residential area (http://madisonfreiburg.org/green/vauban.htm)

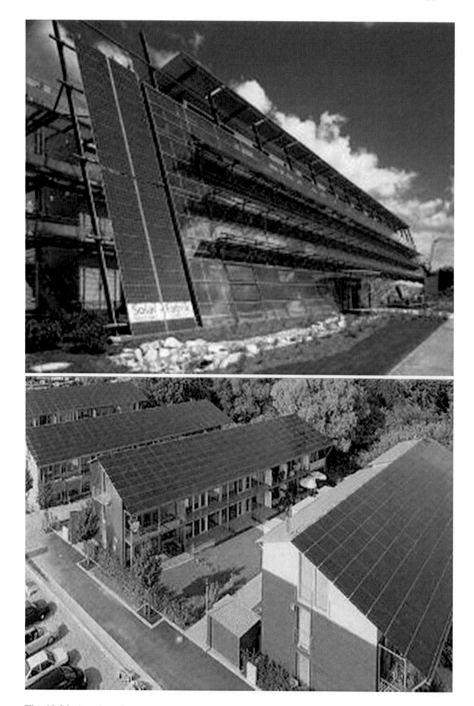

Fig. 10.24 (continued)

- Low-energy building is obligatory in this district; zero-energy and energy-plus building and the application of solar technology are standard for most;
- The rows of old trees were preserved as much as possible;
- The green spaces between the housing rows account for good climatic conditions and provide play areas for children;
- Vegetation-covered flat roofs store rainwater, which is collected and re-used;
- The neighbourhood area is traffic-calmed, with the majority of households not owning a car;
- The residential area has been linked to the city tram system, enabling many people to do without a car, using local public transport or riding their bikes instead.

10.3.1.4 Rieselfeld

The neighbourhood of Rieselfeld combines a positive image, comprehensive public infrastructure and connected neighbourhood life. Civic commitment and proactive cooperation are important in this district. The following characteristics determine the neighbourhood:

- All houses are built as low-energy buildings. In many of them, photovoltaic panels and solar heating utilise the energy of the sun;
- Additional forms of renewable energy utilisation and district heating from a combined heat and power station complement the energy concept;
- A consistent water concept and consideration of climatic aspects;
- green spaces, playgrounds, open areas, bicycle paths and traffic-calmed streets where children are allowed to play are part of the urban design.

Some of the key success-factors in making Freiburg a sustainable city are::

1. An integrated sustainable development vision
2. Support and consensus amongst all political parties on sustainability
3. Engagement and involvement of different stakeholders
4. Participation and commitment of citizens

Sources

www.citiesforpeople.net/in-freiburg-city-its-clean-and-pretty
www.citiesforpeople.net/freiburg-green-city-%E2%80%93-success-factors
www.ice.org.uk/topics/community/Sustainable-Community-Development/Freiburg

10.3.2 Curitiba, Brazil

"Cities are not the problem, they are the solution." is probably the most famous saying of Jaime Lerner, who is the (three time) former mayor of Curitiba and urban planner. Curitiba put this in practice, starting with its first Master Plan, dating from 1965, in which land-use, planning, traffic management and transportation are

Fig. 10.25 Bus lane and bus stop, Curitiba (Photos: Rob Roggema)

integrated and flexibility in regulations *is* maintained to allow for different future development scenarios. One of the central goals was to give all citizens access and this led to the guiding principle that mobility and land use cannot be disassociated. In concrete, the main transport arteries were modified over time to give public transport the highest priority. Each of the five arteries contains one two-way lane devoted exclusively to express buses, safe, reliable, and efficient (Fig. 10.25). About 1,100 buses make 12,500 trips per day, serving 1.3 million passengers. The result of this intensive bus use, Curitiba's gasoline use per capita is 30% below that of eight comparable Brazilian cities it has negligible emissions levels, little congestion, and an extremely pleasant living environment.

Along the five structural arteries high density development is encouraged (Fig. 10.26). This has helped to divert transport movement from the city centre, allowing the transformation of city centre streets into a pedestrian network, which turned out to be a tremendous economic boost. There was much more space available for customers rather than vehicles, the shopping environment was more pleasant, and people had more time to shop when they did not have to drive and park.

The higher densities along central bus-routes go hand in hand with development of a large network of green spaces and parks. The green urban areas in Curitiba are among the largest in the world. The parks improve liveability, but they also are functioning to reduce flooding, rather than canals. Moreover, Curitiba protect local vegetation (mixed subtropical forest), which has been threatened by urban development. It makes sure that the Paraná pine (Araucaria angustifolia) is not felled in public or private parks. In order to protect the local vegetation, the city's Municipal Secretariat of the Environment produces 150,000 endemic cuttings, 16,000 fruit trees and 260,000 flower seeds, at the same time as 350,000 cuttings are nursed in a botanical garden and three greenhouses.

10 The Best City? 235

Fig. 10.26 High rise axes where public transport runs, determining high rise, low rise, ecological structures and recreational lakes (Photos: Rob Roggema)

The population of Curitiba has, among other things, planted 1.5 million trees along the city's highways and byways. Many streets in the city centre have been converted to pedestrian precincts and a 'flower street' is cared for by street children.

Curitiba's 'Open University' provides an education for a modest fee, and the city's inhabitants are taught about environment protection. Clapped out old city buses are used as mobile schools which teach the population about sustainability.

The city has succeeded in introducing a Green Exchange employment program. Low income families living in the favelas, shantytowns out of reach of the city's dustcarts, can exchange their rubbish bags for bus tickets and food. Children can exchange reusable waste with school articles, chocolate, toys and tickets to entertainment events. The project results in less household waste in the streets as well as in sensitive areas such as rivers and parks. In combination with other initiatives, 70% of Curitiba's waste is recycled by the city's inhabitants. The city's recycling of paper alone accounts for the equivalent of 1,200 trees a day.

The long-term consecutive work on the execution of the first Master Plan has made Curitiba one of the most sustainable cities. This can be emphasised by the following:

1. Curitiba has the highest recycling rate in the World – 70%.
2. Curitiba has bus system that is so good that car traffic decreased by 30% while the population trebled in a 20 year period.
3. Curitiba has the largest downtown pedestrian shopping area in the World.
4. Curitiba has built large numbers of beautiful parks to control floods rather than concrete canals. So many that they use sheep to cut the grass as it's cheaper than lawnmowers.
5. Curitiba is a city where 99% of inhabitants want to live. In comparison, 70% of Sao Paolo's residents want to live in Curitiba.
6. Curitiba's average income per person has gone from less than the Brazilian average in the 1970s to 66% greater than the Brazilian average.

Sources
www.citiesforpeople.net/cities/curitiba.html
http://sustainablecities.dk/en/city-projects/cases/curitiba-the-green-capital
www.dismantle.org/curitiba.htm

10.3.3 Malmö, Sweden

Malmö's WesternHarbour (Västra Hamnen) has become the city's most famous district in terms of integrating environment and energy in urban planning. WesternHarbour was an old polluted shipyard, which closed a few decades ago leaving 6,000 people without jobs. But Malmö saw the area not as a problem, but as a solution for the need of a beautiful new part of the city, which would inspire a new environmental, economic and social miracle in Malmö.

The 175 ha artificial island of Västra Hamnen was bought by the Municipality of Malmö in 1996, and the idea was to develop an entirely new eco-district. Over the last 10 years, the municipality has transformed the island from a polluted industrial area to an environment-conscious district with homes, businesses and recreational areas. A fundamentally sustainable approach to planning has been key in the creation of the district with visionary dwelling-types and high-quality permanent housing solutions, architectural diversity and urban spaces where people, aesthetics, ecology and technology merge. It has been highly praised as an exciting, ambitious and thought-provoking success, and the people of Malmö have embraced the district, especially its harbour promenade.

10 The Best City?

Fig. 10.27 Water storage and bicycle paths in Malmo' Western Harbour area (http://spfaust.wordpress.com/2011/02/25/city-of-sustainability-and-renewable-energy/ and http://www.ice.org.uk/topics/community/Sustainable-Community-Development/Malmo)

Sustainability also concerns interaction between the people who live in the area, and objectives have been laid down regarding different forms of ownership in order to reduce the formation of ghettos. Furthermore, design and architecture create aesthetically pleasing urban spaces and attractive places where residents can get together (Fig. 10.27).

This is manifest in such details as protection against the wind and pleasant outdoor areas with a good view and proportions to which residents can relate. In order to ensure a sustainable resource management and recreational and aesthetic values, water in the district flows through an ingenious system of ponds, open channels and moss-covered roofs (Fig. 10.28).

The energy performance of the area is very good as 100% is powered by local renewable energy. Consumption of resources is minimised e.g. by wind turbines, which provide all the electricity and solar panels on the roofs supply a fifth of the heat, the remainder coming from thermal heating and Malmö's existing, super-efficient district heating system. Recyclable and organic materials are sorted and contribute to energy production by the city's biogas plant. The residents are encouraged regularly to check their energy consumption on information panels installed in each home. In addition to this, paths and cycle tracks have been given high-priority as has the use of healthy materials in the dwellings and surroundings.

Additionally, local buses are powered by biogas from residents' waste, there are plenty of green spaces and cars can hardly be seen anywhere.

Interesting enough, not too many people who live there know about the environmental benefits. Malmö took the decision to sell the properties based on them being beautiful and in a beautiful area.

Sources

http://www.citiesforpeople.net/cities/westernharbour.html
http://sustainablecities.dk/en/city-projects/cases/malmo-bo01-an-ecological-city-of-tomorrow
http://www.ice.org.uk/topics/community/Sustainable-Community-Development/Malmo

Fig. 10.28 Ecological water treatment in public space (http://www.ice.org.uk/topics/community/Sustainable-Community-Development/Malmo and http://sustainablecities.dk/en/city-projects/cases/malmo-bo01-an-ecological-city-of-tomorrow)

10.3.4 Masdar City, United Arab Emirates

Masdar – Arabic for *the source* will be the world's first zero-carbon and zero-waste city, developed by Abu Dhabi, the Mubadala Development Company and the British architectural firm Foster and Partners and completion date is expected to be between 2020 and 2025.

10 The Best City?

Fig. 10.29 Masterplan for Masdar City © Foster+Partners (http://www.esri.com/news/arcnews/fall09articles/building-an-oasis.html)

The city will be built 17 km from downtown Abu Dhabi and will cover approximately 6 km^2. It will include several different renewable energy sources. The main source of power will be a 60 MW solar power plant, with plans for more as the city grows. Masdar is planned to be a sustainable, renewable masterpiece of urban design (Fig. 10.29).

It will be car-free and feature, a clean, efficient transport system. Furthermore, use of solar, geothermal and hydrogen energy will power, heat and cool the city and a desalination plant will provide fresh water. Within the limits of Masdar City, the plan is to include water treatment plants, recreational areas and sports facilities, recycling centres, sewage treatment plant, a wind farm, and plantations of different tree species producing biofuels. The ambition is to establish Masdar City as the

Fig. 10.30 Public squares as proposed in Masdar city, shadow combined with solar energy (http://inhabitat.com/foster-partners-carbon-neutral-masdar-city-rises-in-the-desert/)

world's first fully sustainable city, combining traditional planning principles with existing technologies (Fig. 10.30). The city aims to be innovator and pioneer in innovative sustainable technologies for energy, water, and waste management. Masdar City's aim is to show the extent to which cities can be sustainable, while offering world-class living and working environments, thereby changing the global landscape by showing what's possible.

The United Arab Emirates hope that Masdar will not only be a model for sustainable cities and renewable technologies, but also a haven for renewable energy companies, green energy technology research, development and investment, and carbon resource management.

Masdar City is planned to become a high-density, pedestrian-friendly (no cars are allowed) urban area where renewable energy and clean technologies are researched, tested and implemented. The project is inspired by architecture and urban planning of traditional Arab cities. The design incorporates narrow streets, shading of windows, exterior walls, and walkways, thick-walled buildings, and the use of local vegetation. Best practices in traditional urban planning will be combined with advanced practices and technologies in sustainable building. The use of fossil-fuel vehicles is minimised in the city. The design provides the highest-quality living and working environment with the lowest possible carbon footprint.

Upon completion the city will house 40,000 permanent residents and hundreds of businesses.

10.3.4.1 Masdar Plaza

One of the central places in the city will be Masdar Plaza, designed by LAVA (www.l-a-v-a.net/). As many cities across the World have shown, the plaza, forum, or square is the epicentre in the city, the place where life, values, ideals, and visions of the people evolve. The Masdar Plaza is designed as an *iconic beacon* attracting global attention to sustainable technology.

The Plaza is designed to be "the Oasis of the Future": a living, breathing, active and adaptive environment, where social interaction of people occurs, and the use and benefits of sustainable technology are spotlighted. The flexible use of space allows all functions to perform the highest quality of indoor and outdoor comfort and interaction.

Interactive, heat sensitive technology activates low intensity lighting in response to pedestrian traffic and mobile phone usage.

'Petals from Heaven' feature interactive umbrellas that open, provide shade, and capture energy during daylight hours; folding at night to release stored heat (Fig. 10.31).

Solar analysis provides insight into the tuning of facades in order to incorporate an ability to respond to varying sun angles and levels of solar intensity.

The Oasis of the Future has the ability to control ambient temperature at all times of the day. The 'Petals from Heaven' open and close. They protect pedestrians from the sun and capture, store, and release heat. They adjust the angle of shade based on the position of the sun. The heat sensitive lamps adjust the level of lighting to the proximity of pedestrians. The water features ebb and flow based on ground temperatures.

Sources
www.archdaily.com/33587/masdar-sustainable-city-lava/
http://blog.hotelclub.com/sustainable-city-masdar-united-arab-emirates/
http://oilprice.com/Latest-Energy-News/World-News/Masdar-City-The-Most-Sustainable-City-in-the-World-to-Open-in-2025.html
www.careers.ch2m.com/worldwide/en/engineering-projects/masdar.asp

10.4 The *Self-Organising* City

The third option to develop a best city is the opposite of a completely planned city, the city of not-aiming to control and plan and letting the city self-organise the future that over time will appear. There are many examples of these developments. Often, uncontrolled (by government) developments are seen as problematic, such as the Brazilian favela's (Fig. 10.32) and African or Indian slums (Fig. 10.33). Several characteristics, such as criminality or rule less practices may underpin this, but in terms of developing urban patterns these settlements can also be seen as very successful.

Fig. 10.31 Masdar Plaza with its Petals from Heaven (http://thewondrous.com/masdar-city-photos-worlds-first-sustainable-zero-carbon-zero-waste-city/ and http://thewondrous.com/masdar-city-photos-worlds-first-sustainable-zero-carbon-zero-waste-city/)

10 The Best City?

Fig. 10.32 Favela in Rio de Janeiro (Photo: Rob Roggema)

Fig. 10.33 Self-organising slums in Delhi (Photo: Rob Roggema)

10.4.1 Delhi, India

Taking Delhi as an example, and not ignoring some of the serious problems, a couple of principles can be distinguished that illustrate its self-organising capacity.

First of all, the traffic system is, from an outsider perspective, often characterised as chaotic, a mess or worse. However, when driving around for a week in the city three self-organising principles, which clearly provide a logical, understood and safe traffic system:

1. Look only and always at everyone that is in front of you, no matter if it are cars, buses, trucks, bicycles, pedestrians, cows or adverse traffic;
2. Change course slowly, at moderate pace, giving fellow users time to adapt and change their course slowly;
3. When your neighbour comes too close to you, horn.

These simple rules everyone knows, understands and applies determine the functioning of the complex adaptive traffic system in Delhi. It provides a safe system, where hardly any accidents occur.

Secondly, self-organisation is represented in building patterns of seemingly unplanned urban areas. There are clearly no building regulations or design rules put in place, but still, identical patterns develop everywhere. Analysing the route between Delhi and Agra, nearly all settlements along the route developed similar principles. There is a complex system of narrow streets, connected in surprising and unexpected angles, and many different building types and –heights, in the *urban background*. In front of this there is the *business edge* (Fig. 10.34), where all types of business take place. This zone, which changes during the day and night, is always active and lively and contains many different activities (play and leisure, food and drink, mechanics and traffic requirements, etcetera).

The third zone is the *main traffic area*, where fast and slow traffic mingles and the self-organising principles, as described before, dominate.

Together the three zones form the urban fabric, developed according local demands and principles. In its functioning, adaptability and urban patterns it is very sustainable. However, environmentally the sustainability of these areas can be severely questioned.

10.4.2 Cairo, Egypt

Another example of how unplanned self-organisation developed urban pattern can be found in Cairo. During the uprising in the Egyptian capital the Tahrir Square (Fig. 10.35) became the centre-point of activities, unrest and military intervention.

The people gathering on the square were not randomly occupying the space. They deliberately self-organised themselves by creating all necessary functions on the square (Fig. 10.36) that an ordinary society would plan for in a neighbourhood.

"First of all, they took over a nominally public space, which the state wished to exclude them from access to, Tahrir Square. Having taken it over, and affirmed that

Fig. 10.34 Self-organising patterns at the edge of neighbourhoods between Delhi and Agra (Photos: Rob Roggema)

they wouldn't simply go home at the end of the day – something we might want to think about – they saw off wave after wave of assault on the protests, from police and plain clothes thugs. They set up committees to keep watch for government men.... They set up a network of tents for people to sleep in.... There are toilet arrangements – no small logistical matter when there are routinely hundreds of thousands of people occupying the capital's main intersection. They rig up street lamps to provide electricity. They set up garbage collection, medical stops – they occupy a well-known fast food outlet and turn it into somewhere that people shot at or beaten by police can get treated. They set up a city within a city, and collectively coped with many more challenges than the average city would have to face in an average day" (Seymour 2011, cited in Newman 2011).

Fig. 10.35 Tahrir Square, Cairo, Egypt (http://www.ryanlyford.com/myblog/2009/08/ryan-goes-to-frankfurt-cairo/20090705_cairofrankfurt-011/ and http://cairo.neighborhoodr.com/)

Fig. 10.36 Self-organisation patterns resulting from collaborating occupants at Tahrir square (http://www.mumbaishianews.com/2011/02/tahrir-square.html)

10.5 The *Best* City, Delhi Designs for Hobsons Bay

What happens if you place 16 very smart post-docs with different backgrounds in a design charrette environment for an afternoon, and ask them to design their best city? This is what happened during the Indian National Training Program week on Climate Change (Roggema 2012) in Delhi. The students were exposed to four cycles of design work. Three thematic sessions about water management, ecology and coastal protection respectively, were followed by one design session focusing on integration of the themes. With thematic sessions lasting for 25 min and the integration for 45 min the afternoon functioned as e real pressure cooker, which forced the participants to stretch beyond their normal kills. On top of the time constraint came the fact that the chosen case study area was Hobsons Bay, a typical, coastal, Australian suburb in the western part of Melbourne's Metropolitan area, a real novelty for the Indian students. Three very inspiring designs came out of this intensive process: Gagan City, GEO City and Cloud 9.

10.5.1 Gagan City

The design for Gagan City (Fig. 10.37) is based on the underlying principle "Ecologising the urban area and urbanising the water". The water management on land is improved through the creation of new reservoirs in higher lying areas, where the water is stored before it can cause a flood. These new reservoirs are connected with each other and with the existing creeks through a new canal. The canal and the new storage possibilities increase the capacity of the system and make it possible to exchange excess rainwater. Flash flooding is prevented from happening. Excess water is diverted to new harvesting storages where clean water is kept in order to make it available for usage. The locations for these new reservoirs are found in currently unused spaces.

The coastal zone, bound inland by a new withdrawn seawall, is transformed in a broad vegetation zone, creating a new resilient protection area. Where houses exist new wetlands are integrated in current public green space and infrastructure, leading to an urban landscape of houses standing in the middle of a green and wet environment. The parts of the coast without current housing transforms in an area with vegetation and water, allowing for an easy migration of flora and fauna along the coast.

In front of the current coast a new structure is foreseen, providing additional coastal protection: the so-called sky-city. This is a surge barrier, which is open and can be closed when storm surges occur. The elevated bridge/barrier connects Point Cook with Williamstown and consists of a road system and the new sky city. This new city is a high- density city, fully autonomous in energy use: windmills are standing in between the houses on piers and each house is fitted with solar panels.

A final addition in the Port Phillip Bay is projected: the Gagan City International Floating.

10.5.2 GEO City

The basic principle of GEO-City (Fig. 10.38) is to strategically intervene with sustainable measures: an *interventionistic* acupuncture approach. There are several specific interventions determined:

- Related to water management rainwater is stored in different ways: on rooftops, in water tanks and in wetlands. The harvested water will subsequently be used in parks for irrigation purposes;

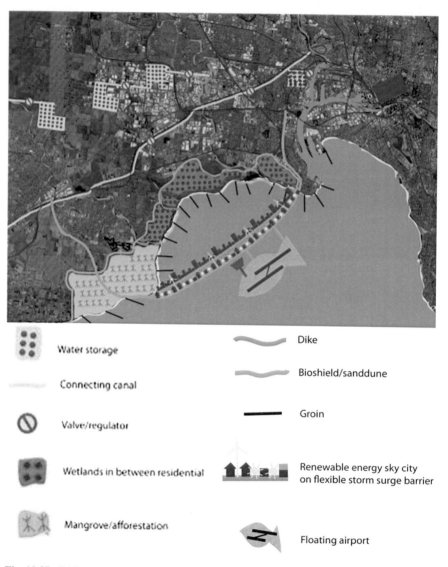

Fig. 10.37 GAGAN City by J. Ansari, G. Girish, V. Mahale and M.P. Prabhakaran

- Industrial buildings will collect rainwater on their rooftops and in water tanks. Besides the water harvesting industrial buildings will be equipped with solar panels to collect energy;
- In parks and public spaces (streets) solar energy will be conserved and along the coastline windmills are projected;
- The residential buildings will incorporate different measures. Some building will have rooftop water storage, while others are provided with hanging gardens or prepared for flood resistance. The choice for a certain measure depends on the

location of the building. When the building can play a role in connecting ecological areas it will contain a hanging garden and where flood risk is apparent the building will be built to deal with that;
- Valuable habitats, currently isolated in the urban patterns are connected through green corridors alongside major roads;
- The coastal protection is arranged for through different proposals: the creation of strategically positioned artificial islands, floating forests, floating houses and an Aqua island, an artificial habitat in the shape of a fish. The coastline is changed into a mangrove forest, beautiful and resilient, but something completely different from Altona's beach shore.

10.5.3 Cloud 9

The emphasis of Cloud 9 (Fig. 10.39) is on creating a compact city, while preserving and creating more space for ecological and climate adaptation measures.

The water management system consists of three parts: the northern part where small dams store excess water in reservoirs, the storage areas in the urban areas and the overflow zone, where rainwater is let in the coastal wetlands. This area protects the land from washing away in the sea.

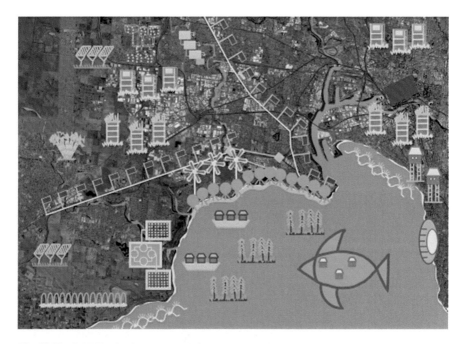

Fig. 10.38 GEOCity by Suman Lata, Priya Dutta, V. Rajesh Kumar and Ratheesh Kumar

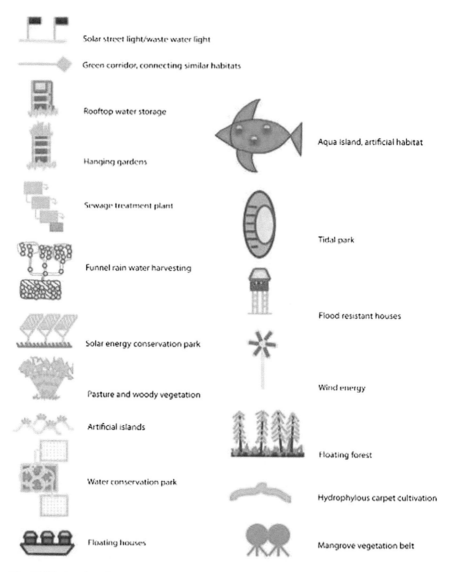

Fig. 10.38 (continued)

The ecosystem provides gradients of vegetation from the remote areas in the northern and western landscapes towards the more artificial habitats closer to the urban areas. The grasslands are ecologically developed through grazing cattle. In the urban environment the density is increased through introducing high-rise buildings, qualifying as green buildings in the existing neighbourhoods. The result of this is that an increased number of people can live in the area, but at the same time the amount of green, ecological, space is growing as well.

10 The Best City?

The land is protected from storm surges through a combination of measures: dense vegetation, wooden barriers, land reclamation, tide resistant barriers, a conventional seawall and floating islands protect the coast.

In conclusion, these three design proposals illustrate the broad variety of measures and interventions that transform the area as it currently is into a climate adaptive area, where biodiversity and ecological qualities are improved, through a variety of measures the coast is protected and a spectrum of water conservation measures the impacts of heavy rainfall are mitigated. Perhaps not all the proposals are immediately realistic, but many of the ideas can be implemented. The designs demonstrate that the existing urban area can transform into an ecological, sustainable and adaptive environment. Setting aside the extreme proposals, such as the floating airport, the designs show that small sustainable interventions and innovative thinking can make an impact, change the image of the urban environment and improve adaptive capacity. These ideas could (only) be developed in an environment of out of the box thinking was enhanced and participants were not hindered by existing policy, regulations or planning habits. The results of this intensive and very short design charrette demonstrated that innovative designs can be created, but is can also be read as a pledge for mutual design sessions, where different countries exchange their cultures and as a pledge for ultra intensive design meetings.

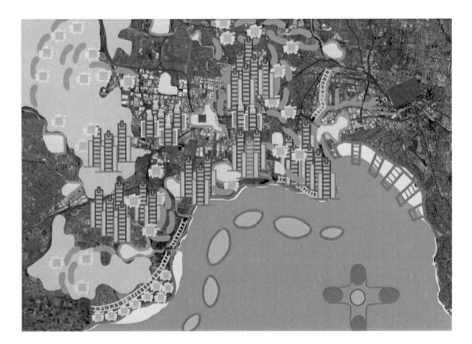

Fig. 10.39 CLOUD9 by A.K. Kanojia, Shakti, Japaechandra and Mohanty

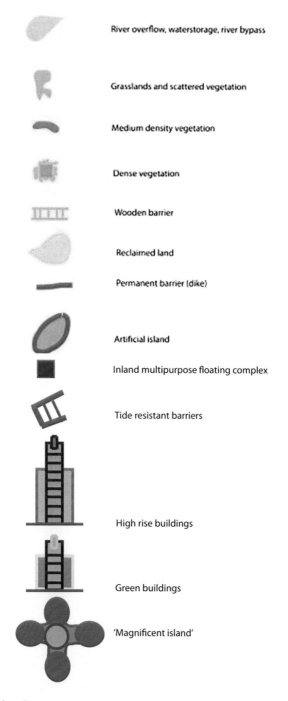

Fig. 10.39 (continued)

10.6 The Art of Designing for Climate Adaptation

According to the Economist Intelligence Unit of the Economist Magazine, Melbourne is currently the most liveable city in the World www.eiu.com/site_info.asp?info_name=The_Global_Liveability_Report&rf=0. However, Melbourne Metropolitan Area is using vast amounts of landscape for residential living in low densities, is oil driven and uses huge amounts of fossil resources (e.g. coal for heating and electricity, oil for transport). The city is under threat of heat, bushfires, flooding as result of sea level rise and flash flooding due to torrential rains, all as result of climate change. It has a troubled public transport system with slow and mostly old trams, with cars obstructing their routes, and a train punctuality of 87%. For cyclists it is a risky ride through the city streets as car drivers don't see them or ignore their presence, often leading to accidents and cyclists being kicked off their bikes. These are some of the backsides of life in the most liveable city in the world. What it tells us is that *the* best city does not exist. There will always be elements that conflict with being the best. And it'll always depend on the standards used.

However, we can learn from examples of the past. Despite the fact that people are very satisfied in the *planned* city, it is too rigid. Despite the fact that the *sustainable* city is ecologically sound, it is too boring. And despite the fact that the *unplanned* city is organised by its inhabitants, it is too environmentally unsustainable. Even designs for the *best* city, despite being prepared for future climate impacts, are, in some parts, too extreme, or simply not good enough. Focusing on the benefits, the good sides of these examples might help to conceive a city that is flexible and able to adapt, and provides a safe and resilient place that challenges its inhabitants to shape their own, sustainable environment. To do some cherry picking from the different cities described before:

1. The learnings of the planned city were that on the one hand side a city is developed, which is highly appreciated by its inhabitants, but it is also seen as too rigid, with over-dimensioned public spaces and buildings. The city shouldn't be planned in each and every detail, but only the strategic interventions should be determined. The rest of the city, then, could be developed over time. Instead of a rigid framework in street patterns, which often made the mistake to be designed for cars instead of people, the spacious green and public spaces should be leading. Additionally, the beauty of exceptional architecture is an elements that deserves to be an element as part of a best city;
2. Learning from the sustainable city the ecological principles can be implemented and will function perfectly. However, in these cities the ideal society seems to live, as perfect citizens obeying the ecological rules, which is sometimes a bit boring and brave. Ecological principles derived from sustainable city and building theory could become an inextricable part of urban design, fully integrated. Moreover, sustainability topics should be very beautiful and could be hyped in order to challenge people to react and adjust. Could these measures provoke and seduce people to cross the line and to change the rules. Then, sustainability has become part of real life and the real city;

3. What has been learned from the self-organising city? There is enormous power in communities and societies to adjust their environment to their liking and requirements. Also, many communities are often constraint by rules, of which the purposes are unclear withstanding developments desired by the community. The backside of rule-less development is their environmental performance, which is often very low. Could self-organising power be combined with clean interventions, which creates a safe and pleasant city, it is beneficial for everyone. Some interventions located in the three zones distinguished in the self-organising city before are:

   In the urban background public beauty-squares can be developed at strategic intersections and other points of importance. These squares can function as place to shelter in case of a disaster and as the provider of clean water, renewable energy and the place where dirt water is cleansed;

   In the business edge beautifully designed delivery points can be introduced. Here clean water and energy can be retrieved and garbage can be brought in. When there are plenty of these points created they are easy accessible for everyone;

   The main traffic zone currently a main polluter and this can be prevented if clean fuel, electricity and bio-fuels, is cheap and easy to get in frequently available filling stations.

4. The main learning from the Delhi best cities designs is that unexpected solutions and designs can be developed in a very short period, by people who are not constraint by existing rules and policies and lead to innovations. Not every single proposal is easily implemented, but the majority of ideas can and is capable of refreshing the minds of local policy makers.

In conclusion, the components of the art of designing for climate adaptation can be summarised in the following six actions:

1. **Analyse**: The design process starts with an analysis of the existing and logical developed networks. The networks determine future growth directions and potential new combinations of functions. Additionally the existing functional lay out of an area supports the network analysis;
2. **Focus**: Secondly, the main nodes in the spatial field are identified. These points are highly likely the places where new developments ignite. Here, the attractors can be found and 'it' all happens. When an intervention is done in these locations the entire spatial patterns will be influenced. It can be expected that spatial transformations will start from these nodes;
3. **Free**: Allow for a free development in the majority of the urban landscape. When rules are too rigid self-organising power is bound and will not create a strong resilient urban system;
4. **Plan**: Put a strategy in place to plan the unplanned. A strategy for including climate adaptation measures consists of three types of measures: (1) start with the inevitable (decisions already taken and cannot be withdrawn). This leads to the starting point, drawn on a map; (2) add no regret measures (being the strategic points) and other measures that are always beneficial, no matter what the future might bring, such as extreme weather events; and (3) add space for reservations to accommodate the unknown;

10 The Best City?

Fig. 10.40 A tale of two cities (Billard Leece Partnership Oty Ltd.)

5. **Release**: When major interventions are identified and implemented the control over execution, the spatial layout or the future needs to be released. A new spatial layout can only evolve when there is no control over the long-term and in great detail;
6. **Check-up**: The intended future, as determined by the strategic interventions in key nodes and the enhancement of self-organising developments in other areas needs a regular check-up. It could be that new key nodes and networks are developed over time. This may offer an opportunity to create a new intervention. When these regular check-ups need to take place is unidentified. This depends on the pace of general changes (in economy, demography, climate, culture, politics).

The best city doesn't exist. However, designers, planners and futurists will continue to design their Utopias. As the Tale of Two Cities (Fig. 10.40) illustrates, the fantasy of Modern Utopians is rich. This second cities assesses all building and design proposals on their sustainability, providing information on a continuous renewal and recycling of the city. This example is one of the 17 new Australian urbanisms, that reinvent the city and create a better, more sustainable world (Gollings et al. 2010).

References

DiChristina M, with editors (2011) Street-Savvy, Meeting the biggest challenges starts with the city. Scientific American, September 2011, pp 26–75
Eaton R (2003) Ideal cities: Utopianism and the (un)built environment. Thames & Hudson, London
Gemeente Almere (2009) Summary draft structural vision Almere 2.0. Almere: gemeente Almere. http://english.almere.nl/local_government/almere_2.0
Gollings J, Rijavec I, Bremner C (2010) Now and when Australian urbanism. Australian Pavilion 12th International Architecture Exhibition la Biennale di Venezia 2010. The Australian Institute of Architects, Canberra
Kassenaar B (1994) Ideeën voor een duurzame stad; het geval Amsterdam. Milieudienst, Amsterdam
Logan GM, Adams RM (1989) Thomas More: Utopia (Cambridge, Cambridge University Press).
Neuwirth R (2011) Global bazaar. Scientific American, September 2011, pp 42–49
Newman S (2011) Postanarchism and space: revolutionary fantasies and autonomous zones. Plan Theory 10(4):344–365
Ratti C, Townsend A (2011) The social nexus. Scientific American, September 2011, pp 30–35
Roggema R (2012) Design of a climate adaptive Hobsons Bay. Hands-on training. In: Proceedings "Climate change and geospatial technology". National Level Training Program. CSIR-NISCAIR, Delhi, India
Seymour R (2011) Towards a new model commune. Lenin's tomb blog: http://leninology.blogspot.com/2011/03/towards-new-model-commune.html. Accessed 8 Nov 2011
Stassen B (2001) Bedacht en gebouwd, 25 jaar Almere-Stad. Almere, Gemeente Almere, Dienst Stedelijke Ontwikkeling

Websites

http://blog.hotelclub.com/sustainable-city-masdar-united-arab-emirates/. Accessed 20 Feb 2012
http://oilprice.com/Latest-Energy-News/World-News/Masdar-City-The-Most-Sustainable-City-in-the-World-to-Open-in-2025.html. Accessed 20 Feb 2012
http://sustainablecities.dk/en/city-projects/cases/curitiba-the-green-capital. Accessed 20 Feb 2012

http://sustainablecities.dk/en/city-projects/cases/malmo-bo01-an-ecological-city-of-tomorrow. Accessed 20 Feb 2012
www.archdaily.com/33587/masdar-sustainable-city-lava/. Accessed 20 Feb 2012
www.careers.ch2m.com/worldwide/en/engineering-projects/masdar.asp. Accessed 20 Feb 2012
www.citiesforpeople.net/cities/curitiba.html. Accessed 20 Feb 2012
www.citiesforpeople.net/cities/westernharbour.html. Accessed 20 Feb 2012
www.citiesforpeople.net/in-freiburg-city-its-clean-and-pretty. Accessed 20 Feb 2012
www.citiesforpeople.net/freiburg-green-city-%E2%80%93-success-factors. Accessed 20 Feb 2012
www.dismantle.org/curitiba.htm. Accessed 20 Feb 2012
www.eiu.com/site_info.asp?info_name=The_Global_Liveability_Report&rf=0. Accessed 5 May 2012
www.ice.org.uk/topics/community/Sustainable-Community-Development/Freiburg. Accessed 20 Feb 2012
www.ice.org.uk/topics/community/Sustainable-Community-Development/Malmo. Accessed 20 Feb 2012
www.l-a-v-a.net/. Accessed 2 May 2012

Subject Index

A
A-apostrophe pathway, 87, 88
A-Catastrophe, 46
Adaptation, 46, 49, 126. *See also* Sector-based adaptation
Adaptive capacity, 39, 45, 127, 131–133, 142, 144, 146, 149, 154, 187, 251
Advanced transition, 74–75. *See also* Transition
Afghan stability operation, networks used in, 93, 97
Aggregation, in spatial system, 15–17
Agonism, 121, 122
Ajax, 34, 35, 39
Almere (Netherlands, The)
 Dutch climate, 224–225
 houses, 223–225
 iconic buildings, 224
 layout, 223
 streetscape, 225
 urban districts, 222
Archigrams, 214
Art of designing, for climate adaptation, 253–256
Assembly building (Chandigarh), 213
Attractor, 46, 126
Australia. *See also* Commonwealth of Australia; Melbourne Metropolitan Area
 Canberra, 226–228
 design charrette processes, 159
 urban design, 7

B
Backcasting, 85
Backtracking, 85
Badenova Stadium, 231
Baronielaan plan (Breda), 27, 28
Bendigo, bushfire resilient landscape of, 183
 freedom to emerge, 186–187
 whole-system level, intervention at, 184–186
Best city, 207
 climate adaptation, art of designing for, 253–256
 Delhi designs, for Hobsons Bay, 246
  Cloud 9, 249–252
  Gagan City, 247
  GEO City, 247–249
 planned city
  Almere, 222–225
  Brasilia, 219–222
  Canberra, 226–228
  Chandigarh, 216–219
  communalities, 228–229
 self-organising city, 241, 243–246
  Cairo, 244–246
  Delhi, 244, 245
 sustainable city, 229
  Curitiba, 233–236
  Freiburg, 229–233
  Malmö, 236–238
  Masdar City, 238–242
Bifurcation points, 126–129, 144
Bijlmer-area (Amsterdam-southeast), 29
Biological network, 93, 95
Biomass Combined Heat and Power (Bio-CHP) installations, 111
Biomimicry, 199
Black Saturday bushfires, 178, 183
Blauwe Stad, 172–174
B-minus, 81–85, 87, 88

"Bocage" landscape, 55
'Both ends' society, emergence of, 34
Bottom-up approach, 3–4
Brasilia (Brazil)
 central axis, 221
 climate, 221
 iconic buildings, 221, 222
 Masterplan, 219, 220
 neighbourhoods, 219, 220
Brazil
 Brasilia, 219–222
 Curitiba, 233–236
Breugels, Pieter, 214
Breukelen The Van der Valk Hotel, 211
Broadacre City, 214
Brundtland Committee, 197
Bushfire resilient landscape
 of Bendigo, 183–187
 of Murrindindi, 177–183
Business-as-Usual scenario, 6, 8, 11
Butterfly Effect, 45

C

Cairo (Egypt), 244–246
Canberra (Australia)
 climate, 227
 iconic buildings, 227
 layout, 226
 main axis, 226
 neighbourhoods, 227, 228
 streetscape, 228
CAS. *See* Complex adaptive systems (CAS)
CBD, 6, 17, 18
Centralised network, 32, 33
Centre for Development of Creative Thinking (COCD), 142, 161, 162
Chalk drawings, 214
Chandigarh (India)
 Capital Complex, 217
 city centre, 218
 climate, 218
 iconic buildings, 219
 Masterplan, 216
 street in, 217
Chaos, 4
Charrette, definition of, 152–153
Chaux (Ledoux), 214
Citroens, Paul, 214
City. *See also* Best city
 as complex systems, 129–130
 and energy, 202
  cities rather than buildings, 202
  energy and climate, 202
  excess and shortage, 202–203
 as organisms, 195, 198
  background, 196–198
  individual and collective life, 199–202
  life, definition of, 199
City Hall, 231
"City-level development strategy,", 48
Climate adaptation, 1, 50, 118.
 See also Climate change
 art of designing for, 253–256
 current planning practice, 5
  Greater Groningen Area, 8, 10–11
  Melbourne Metropolitan Area, 6–9
  prevailing plan *vs.* climate adaptation requirements, 12–14
 design and policy, 2–4
 layer theory, 144
 malintegration, 2
 in spatial planning, 2
 spatial system, as CAS, 15–18
 wicked problems, 18–21
Climate change, 26, 45. *See also* Climate adaptation
 effects, 58
 and energy, 5, 202
 planning strategies for, 53
  flexible adaptation planning, 58–60
  integrated adaptation planning, 57–58
  mitigation, planning for, 54–55
  sector-based adaptation, 55–57
 and spatial planning, 39, 48–50
Climate design(s)
 and networks, 91
  intensities, explorations on, 98–100
  network theory, 92–98
  Peat Colonies, application in, 101–113
Climate landscapes, 173
 Bendigo, bushfire resilient landscape of, 183–187
 floodable landscape, 174–177
 Murrindindi, bushfire resilient landscape of, 177–183
Cloud 9
 design proposals, 251
 ecosystem, 250
 land, 251
 water management system, 249
Clustering, 27, 86, 94, 98, 108
COCD. *See* Centre for Development of Creative Thinking (COCD)
Collective life, 199
 cybernetic systems, 201
 energy, 199–200

Subject Index

evolution, 201
order, 199
self-perpetuation, 201
separation, 200–201
symbiosis, 201–202
Commonwealth of Australia, 19–21
Communalities, 228–229
Communicative planning, 58, 61, 118, 122
Complex adaptive systems (CAS), 45–46, 69, 119, 126
  Gaia as, 16
  planning process, 51
  properties
    aggregation, 15
    diversity, 15
    flows, 16
    nonlinearity, 15
  in spatial planning, 50–53
  spatial system as, 15–18
Complexity
  exploring, 126–133
  in planning, 130–131
Complexity theory, 126–129
  concepts of thought, 45–48
  origins, 44–45
Complex large-scale systems, 3, 4
Complex system(s), 15
  behaviour, 129
  cities as, 129–130
  evolution, 46, 47
  framework, 52, 53
  self-organisation, 126
  in spatial planning, 52
Computer network, 93, 94
'Continuous Monument,', 93, 214
'Creating Sustainable Cities,', 198
Critical systems, 15
Cruijff, Johan, 34, 35
Cultural dominated approach, 5
Curitiba (Brazil)
  bus lane and bus stop, 234
  central goal, 234
  Green Exchange employment program, 235
  green urban areas, 234
  high density development, 234
  Master Plan, 233, 236
  'Open University,', 235
  parks, 234
  population, 235
Current networks, 92
Cybernetic systems, 201
Cyber networks. *See* Social networks

D
Dantes Divina Commedia, 73, 74
Decentralised network, 32, 33
Delhi (India), 244, 245
Delhi designs, for Hobsons Bay, 246
  Cloud 9, 249–252
  Gagan City, 247
  GEO City, 247–249
Democratisation, of planning process, 30
Design
  charrettes, 142
    Groningen charrettes, 155–157
    involvement through design, 152–155
    regular planning processes, 150–152
    success factors, 159–160
    Victorian design charrettes, 157–159
  framing and policymaking, 2–4
  involvement, 152–155
  problems, 18–19
  and spatial planning, 151
"Dilemmas in a General Theory of Planning,", 118
Directed network, 95
Disch, Rolf, 231
Disconnected waves, 76, 77
Distributed networks, 33
Disturbed reactive environment, 26, 29
Diversity, in spatial system, 15, 17–18
Docters van Leeuwen, Arthur, 35
Dutch Delta Committee, 177
Dutch planning framework, 11–12
Dutch urban design, 10
Dynamic system, 46, 53

E
Early signals, 84
  starting points, creation of, 86–87
  warning signals, 85–86
Earthquakes, 37, 38
Earth system, 49, 69
Earth System Models (ESM), 37
Economic Crisis, The, 26
Economic network, 93, 94
"The edge of chaos,", 46
Education law (1904), 27
Eemsdelta region, 174–176, 187
Egypt. *See* Cairo
Energy, 199–200
  and cities, 202
    cities rather than buildings, 202
    energy and climate, 202
    excess and shortage, 202–203
  Peat Colonies, network in, 102–103, 105, 106

Energy Potential Mapping (EPM), 197
Environment. *See Specific* Environment
EPM. *See* Energy Potential Mapping (EPM)
ESM. *See* Earth System Models (ESM)
Etihad stadium (Melbourne), 213, 214
Exchanging and cascading energy, 204
Explicit knowledge *vs.* tacit knowledge, 153–154

F
Far-future networks, 92
Favela (Rio de Janeiro), 243
Ferris, Hugh, 214
Fitness landscape, 126, 127, 132, 133
Flexible adaptation planning, 58–60
Floodable landscape, 54, 58, 59, 168, 174–175
　freedom to emerge, 177
　whole-system level, intervention at, 176–177
Floods, 22, 38, 44, 48, 49, 55, 57, 59, 60, 145, 179, 182, 189, 190, 236
Flows, in spatial system, 16, 18
Focal points, 144, 146, 148
Forecasting, 36, 84, 85
Framing design, and policymaking, 2–4
Freiburg (Germany), 229
　City centre, 230
　Rieselfeld, 233
　solar city, 230–232
　traffic, 230
　Vauban Quarter, 231, 233

G
Gagan City, 247
Gaia, as CAS, 15, 16
GEO City, 247–249
Germany. *See* Freiburg
Global energy supply, and climate change, 5
Governance, 50, 154
'Grand Projets,', 27
Great Dividing Range, 178–179
Greater Groningen Area, 8, 10–11
Green Exchange employment program, 235
Greenfield developments, 6
Grimsvatn volcano, eruption of, 37
Groningen charrettes, 155–157
Groningen Museum, 172, 173
Groningen province, application in, 146–151
Growth
　cycles, 77, 78
　phases, 76
　population, 6, 26
　and transition, 77

H
Haagse Beemden (Breda), 30
The Hague area, 27
Heavenly city, medieval miniatures of, 214
Hobsons Bay, Delhi designs for, 246
　Cloud 9, 249–252
　Gagan City, 247
　GEO City, 247–249
3-horizons model, 70, 71
Hotspot Climate-proof Groningen project, 155
Housing law (1901), 27
Howard, Ebenezer, 214
Human genes, 93, 97
Hurricane Katrina, 37

I
'Ideal Cities: Utopianism and the (Un)Built Environment,', 214
Idea-map, for adaptive Groningen, 13–14
Incrementality, 13, 69, 121
India
　Chandigarh, 216–219
　Delhi, 244, 245
Indian National Training Program week on Climate Change, 246
Individual and collective life, 199
　cybernetic systems, 201
　energy, 199–200
　evolution, 201
　order, 199
　self-perpetuation, 201
　separation, 200–201
　symbiosis, 201–202
Industrialisation, 27
'Instant City,', 214
Integrated adaptation planning, 57–58
Integrated climate design, for Peat Colonies, 112–113
Intensities, explorations on, 98–100
Interconnectedness, 4
Intergovernmental Panel on Climate Change
Internet-economy, 32–34
Interventionistic acupuncture approach, 247
Island Utopia (Thomas More), 215

J
Jeanneret, Charles-Édouard. *See* Le Corbusier

K

Kinglake-Murrindindi region, 177–179
Knowledge creation, 153–154
Koolhaas, Rem, 214

L

Landscape 2.0, 33
Landscape Architecture, themes for, 18–19
Large-scale systems, 2
Layer approach, 143–145
Le Corbusier, 212
Lerner, Jaime, 233
Life
 collective and individual, 199–202
 definition, 199
Living Lab, 160
'Lonelycolony' model, 108, 110–111
Long-term developments, 5
Lorenz, Edward N., 44, 45
Lotus Temple (Delhi), 209, 210
'Luilekkerland,', 214

M

Maagdenhuis-riots (Amsterdam), 29
Malintegration, of climate change, 2
Malmö (Sweden), 236–238
Masdar City (United Arab Emirates), 238–242
Masdar Plaza, 241, 242
Medieval miniatures, of heavenly city, 214
Melbourne, Victorian housing in, 212, 213
Melbourne Metropolitan Area, 6–9
'Metropolis collage,', 214
Middle East, 26
Mitigation, 49, 54–55
Mitterand (President), 27
Modernism, 219
Moment uncertainty, 37–39
Mono-rationality, 124, 125
More, Thomas, 214
Murrindindi, bushfire resilient
 landscape of, 177
 freedom to emerge, 182–183
 whole-system level, intervention at, 180–182

N

National Laws, 27
Natural disasters, 37–38
Natural resources, 145, 150
Near-future networks, 92

Netherlands, The, 8, 10, 27, 51, 196
 Almere, 222–225
 climate adaptation, 13
 design charrette processes, 159
 spatial planning, 13
Networks, 144. *See also Specific* Networks
 characteristics, 86–87
 and climate design, 91
  intensities, explorations on, 98–100
  network theory, 92–98
  Peat Colonies, application in, 101–113
Network theory, 86, 92–98
Neural network, 93, 95
Neutrality, 121
New Babylon, 214
New Stepped Strategy, 203
Newton, 45, 70, 71
Niche innovations, 77, 78, 84
Nieuwmarkt (Amsterdam), 30
Nieuw-West (Amsterdam-west), 29
Non-linear dynamic systems, 46
Nonlinearity, in spatial system, 15, 17

O

Oasis of the Future, 241
Occupation patterns, emergence of, 145
Opera House (Sydney), 209, 210
Order, 4
Organisational environments, 26
Organisms, city as, 195, 198
 background, 196
 individual and collective life,
  features of, 199
 life, definition of, 199
 'The Origin of Life,', 198

P

Parijs, Le Corbusiers plans for, 214
Path dependency, 131
Peat Colonies, application in, 101, 102
 climate designs for, 108
  integrated design, 112–113
  'Lonelycolony' model, 108, 110–111
  'Peatcometro' model, 111–112
 energy network, 102–103, 105, 106
 transport network, 107–109
 water network, 101, 103, 104
Periphery network, 94
Phalanstere of Fourier, The, 214
Pillai, Sundaresan, 208
Placid clustered environment, 26, 27
Placid randomised environment, 26

Planned city
    Almere, 222–225
    Brasilia, 219–222
    Canberra, 226–228
    Chandigarh, 216–219
    communalities, 228–229
Planning. *See also* Spatial planning
    climate change, strategies for, 53
        flexible adaptation planning, 58–60
        integrated adaptation planning, 57–58
        mitigation, planning for, 54–55
        sector-based adaptation, 55–57
    complexity in, 130–131
    and design, 2–4
    horizon, 5
    journals
        dynamic division, 124–125
        integration, 124
        intervention, 125
        mono-rationality, 125
        paradigm shift, 125
        wicked problems, 125–126
    paradigms, 120–124
    practice
        Greater Groningen Area, 8, 10–11
        Melbourne Metropolitan Area, 6–9
        prevailing plan *vs.* climate adaptation requirements, 12–14
    process, 51, 52
    unsafe, 123, 124, 126
    urban, 30
Plasticine, 158–160, 164
Political timeframe, 5
Population growth, 6, 26
Port Sunlight, 214
Positivism, 121
Post-anarchism, 126
Post-positivism, 121–122
Post-structuralism, 121–122
Power, 29, 30
Prevailing regional plan *vs.* climate adaptation requirements, 12–14
Project management, role of, 48

R
Random network, 94
REAP. *See* Rotterdam Energy Approach and Planning (REAP)
Recognized ignorance uncertainty, 36
Reflexive planning, 121, 123
Renewables, 197

Resilience, 68–69
    approach, 38
    development, 36
    of spatial system, 38
Rhythm, 143, 200
Rieselfeld, 233
Robust adaptation decisions, 36, 37
Robust network, 95
Rotterdam Energy Approach and Planning (REAP), 203–204

S
SASBE. *See* Smart and Sustainable Built Environments (SASBE)
Scale-free network, 93, 94
Scenario uncertainty, 36
Sea level rise, 13, 36, 44, 48, 55, 57, 59, 169, 174, 175, 177, 189, 225, 253
Sea palace (Amsterdam), 210, 212
Sector-based adaptation, 55–57
Self-organisation, 35, 39, 46, 126
Self-organising city, 241, 243
    Cairo, 244–246
    Delhi, 244, 245
Self-organising network, 93, 94
Self-perpetuation, 201
Slow pace dynamic layer, 145
Slow pace transition, 74–75.
    See also Transition
Smart and Sustainable Built Environments (SASBE), 160
Social networks, 93, 94, 215
Socio-ecological systems, 68–69
Socio-technological transitions, 68.
    See also Transition
Space Frame Cities (Friedman), 214
Spatial elements, definition of, 69
Spatial planning, 27, 29, 34, 35, 39, 49–50.
    See also Planning
    CAS, 50–53
    changes in, 69–70
    and climate change, 2, 19, 39, 48–50
    definition, 118
    and design, 2–4, 151
        cultural dominated approach, 5
    discourses, 118
    in the Netherlands, 13
    paradigms, 119
    process, 51
    Victorian context for, 6
    and wicked problems, 19

Subject Index

Spatial system
  aggregation in, 15–17
  as CAS, 15–18
  diversity in, 15
  Earth system, 69
  flows in, 16
  nonlinearity in, 15
  resilience, 38, 69
Squatter-riots, 29
Statistical uncertainty. *See* Value uncertainty
'Steer the swarm' strategy, 170–171
Structural uncertainty, 36
Successive Limited method, 121
Superstudio, 214
Sustainable city, 229
  Curitiba, 233–236
  Freiburg, 229–233
  Malmö, 236–238
  Masdar City, 238–242
Sustainable community, 153
Sustainable development, 201
Swarm, 38, 40
Swarming landscapes, 167
  climate landscapes, 173
    Bendigo, bushfire resilient landscape of, 183–187
    floodable landscape, 174–177
    Murrindindi, bushfire resilient landscape of, 177–183
  interventions, 171
    Blauwe Stad, 172–174
    Groningen Museum, 172, 173
  strategies, 168
    impulses, 168–170
    'steer the swarm' strategy, 170–171
  swarm planning experiment, 187
    destructive mob-elections, 191–192
    on the move, 189–190
    sustainable emergence, 190–191
    time incongruence, 188–189
Swarm planning, 53, 58–60, 131, 141, 205
  design charrettes, 150
    Groningen charrettes, 155–157
    involvement through design, 152–155
    success factors, 159–160
    Victorian design charrettes, 157–159
  in Eemsdelta area, 176
  experiment, 160–163, 187
    destructive mob-elections, 191–192
    on the move, 189–190
    sustainable emergence, 190–191
    time incongruence, 188–189
  framework, 143
    Groningen province, application in, 146–151
    layer approach, 143–145
    usage, 145–146
  freedom to emerge, 132–133
  intervention, 132
  theory (*See* Swarm planning theory)
  whole and the parts, 142–143
Swarm planning theory, 117
  approach, 119–120
  complexity, exploring, 126
    complexity theory, 126–129
    complex systems, cities as, 129–130
    planning, complexity in, 130–131
    swarm planning, 131–133
  current planning paradigms, 120
    planning journals, review of, 124–126
    selection, 120–124
  problem statement, 119
Sweden. *See* Malmö
Symbiosis, 201–202

T

Tacit knowledge *vs.* explicit knowledge, 153–154
Tackling wicked problems, 20–21
Tahrir Square, 244, 246
"Three-Bodies" problem, 45
Time-horizons, 5, 92
Tipping points, 58, 120, 126, 127, 132, 134
Top-down approach, 2–3, 154
Transformation(s), 33–34, 75, 154
  A-apostrophe, 80–81
  B-minus, 80–81
  disconnected waves, 76, 77
  growth
    cycles, 77, 78
    phases, 76
    and transition, 77
  multi-level perspective, interaction between, 79
  novelties, uptake of, 79
  phases, 76
  process of change, 78
Transition
  advanced, 74–75
  definition, 70
  and growth, 77
  phases, 70–74
  slow pace, 74–75
  three horizons of change, 70, 71

Transport network, in Peat Colonies, 107–109
Turbulence, 25
  beginning twentieth century–mid twentieth century, 27
  beyond turbulence, 32
    Internet-economy, 32–34
  eighties and nineties, 30–31
  end nineteenth–beginning twentieth century, 26–28
  nineties and early twenty-first century, 31–32
  sixties and seventies, 29
Turbulent environment, 26, 31, 39, 48, 168

U
Uncertainty, 34–37
  moment, 37–39
  recognized ignorance, 36
  scenario, 36
  structural, 36
  value, 36
United Arab Emirates. *See* Masdar City
United Nations Climate Change Conference (Copenhagen)
Unplanned space, 144–145, 149
Unsafe planning, 123, 124, 126
Urban development, 6, 26
Urban Growth Boundary, 6, 8
Urban heat island effect, 55, 57
Urban planning, 30
Urban system, 142

V
Value uncertainty, 36
Varna, floods, 58, 60
Vauban Quarter, 231, 233

Vervoorn, Hans, 73
Victorian design charrettes, 157–159
Victorian housing, in Melbourne, 212, 213
Victorian planning framework, 6, 11–12
Volcanoes, 37, 38
'The voluntary prisoners of Architecture,', 214
Vortical environment, 26, 31
Vouga region (Portugal), 55, 57
Vreewijk (Rotterdam), 27
Vulnerability, 2, 49, 55, 184

W
Water network, in Peat Colonies, 101, 103, 104
Whole-system level, intervention at
  Bendigo, bushfire resilient landscape of, 184–186
  floodable landscape, 176–177
  Murrindindi, bushfire resilient landscape of, 180–182
Wicked problems
  climate change, 58
    characteristics, 18
    Landscape Architecture, themes for, 18–19
    spatial planning and climate change, 19
  tackling, 20–21
  definition, 118
Window of opportunity, 76, 78, 84
'Windows of Groningen,', 169, 170
'Wishing-cards,', 155
World Commission on Environment and Development, 201
World Wide Web representations, 93, 95, 96
Wright, Frank Lloyd, 214